OpenStreetMap

Be your own Cartographer

Jonathan Bennett

BIRMINGHAM - MUMBAI

OpenStreetMap

First published: September 2010

Production Reference: 1170910

Published by Packt Publishing Ltd.
32 Lincoln Road
Olton
Birmingham, B27 6PA, UK.

ISBN 978-1-847197-50-4

www.packtpub.com

Cover Image by John M. Quick (john.m.quick@gmail.com)

Credits

Author
Jonathan Bennett

Reviewers
Steve Coast

Richard Fairhurst

Acquisition Editor
Sarah Cullington

Development Editor
Chaitanya Apte

Technical Editor
Roger D'souza

Copy Editor
Neha Shetty

Indexer
Tejal Daruwale

Editorial Team Leader
Mithun Sehgal

Project Team Leader
Priya Mukherji

Project Coordinator
Prasad Rai

Proofreader
Clyde Jenkins

Production Coordinator
Aparna Bhagat

Cover Work
Aparna Bhagat

About the Author

Jonathan Bennett is a British journalist, writer, and developer. He has been involved in the OpenStreetMap project since 2006, and is a member of the OpenStreetMap Foundation. He has written for print and online technical publications including PC Magazine, ZDNet, CNET, and has appeared on television and radio as a technology commentator. He has an extensive collection of out-of-date printed maps.

To Katie, Matthew, Joseph and Rosie. Thank you for putting up with the little excursion every journey with me involves.

I would like to thank Steve Coast, Richard Fairhurst, Andy "Gravitystorm" Allan, Grant "Firefishy" Slater, Tom Hughes, and Matt "Zerebubuth" Amos for help while writing this book, and anyone who's ever added something to OpenStreetMap, from a single postbox to entire road networks.

About the Reviewers

Steve Coast has worked in a variety of heavy lifting computing applications, from computational geometry to artificial life, and mapping applications. Steve lives in Denver, Colorado, and enjoys hangliding, Mongolian throat singing, and soccer.

As founder of OpenStreetMap, the free wiki world map, Steve undertakes a wide variety of roles within the project, from creating the API to his many public speaking appearances around the world. Steve is also chairman of the OpenStreetMap Foundation, a not-for-profit that supports and guides the project.

> I'd like to thank the thousands of OpenStreetMap contributors around the world for making the best map possible.

Richard Fairhurst is a journalist, cartographer, and programmer who has been involved with OpenStreetMap since the very first days of the project. He is best known for the Potlatch online map editor, but his favorite OSM activity is strapping a GPS to the handlebars of his trusty hybrid bike and going out to map a National Cycle Network route.

His day job is editing Waterways World magazine. In his spare time, as well as mapping, he is a volunteer Sustrans cycle ranger and plays the church organ. He lives in a small Oxfordshire town, which is reputed to have the highest concentration of OSM users anywhere in the world.

Table of Contents

Preface

This book will allow you to take control of your own maps and use them smoothly. This book introduces the reader to the OpenStreetMap project and shows you how to participate in the project, and make use of the data it provides. No prior knowledge of the project is assumed, and technical details are kept to a minimum.

What this book covers

Chapter 1, Making a Free, Editable Map of the World explains how and why the project was started, how freely redistributable geographic data is different from maps that are merely free of charge to use, and some of the things the project has already achieved.

Chapter 2, Getting Started at openstreetmap.org covers the main features of openstreetmap.org and how to create and personalize an account on it. It explains how to use the documentation in the OpenStreetMap wiki, and how to communicate with the OpenStreetMap community.

Chapter 3, Gathering Data using GPS looks at the tools and techniques used by the OpenStreetMap community to gather data using GPS and upload it to the website. It also explains some basic surveying techniques.

Chapter 4, How OpenStreetMap Records Geographical Features covers the types of data you can record, such as nodes, ways, and relations. It also looks at the tagging system and how the community uses and manages it.

Chapter 5, OpenStreetMap's Editing Applications looks at the basic operation of OpenStreetMap's three most popular editors, including Potlatch, JOSM, and Merkaartor.

Chapter 6, Mapping and Editing Techniques explains how to draw features based on GPS traces, how to tag them, and how to find tags that mappers have used, but not documented.

Chapter 7, Checking OpenStreetMap Data for Problems explains how to find the cause of any problems you're having with OpenStreetMap data using the data inspection tools on openstreetmap.org, the NoName layer, ITOWorld OSM Mapper, and Geofabrik's OSM Inspector.

Chapter 8, Producing Customised Maps explains how to create maps using the standard renderings on openstreetmap.org, using a standalone rendering application for Windows and Kosmos, and in Scalable Vector Graphics (SVG) format using Osmarender.

Chapter 9, Getting Raw OpenStreetMap Data discusses ways of accessing the data in the OpenStreetMap database, such as Planet files, the main OpenStreetMap API, and the Extended API (XAPI).

Chapter 10, Manipulating OpenStreetMap Data using Osmosis looks at a tool that is used heavily within OpenStreetMap to manipulate data from planet files or extracts, or databases containing OpenStreetMap data, called Osmosis.

Chapter 11, OpenStreetMap's Future explains some of the changes being developed by the coders and mappers working on OSM, and how they'll affect users of the data.

Who this book is for

This book would be the perfect aid for geographic-information professionals interested in using this data in their work, and web designers and developers who want to include mapping in their sites, and want a distinctive style. This book is for you if you have a need to use maps and geographic data for work or leisure, and want accurate, up-to-date maps showing the information you're interested in, without details you don't need. If you want to use maps for navigation, and want more or less detail than traditional printed maps give, this book would be perfect for you.

Conventions

In this book, you will find a number of styles of text that distinguish between different kinds of information. Here are some examples of these styles, and an explanation of their meaning.

Code words in text are shown as follows: "The `role` attribute is a simple string whose values and significance is defined by the type of the relation itself."

A block of code is set as follows:

```
<osm version="0.6" generator="OpenStreetMap server">
  <node id="483034256" lat="55.9458449" lon="-3.2035477" version="1"
    changeset="2369219" user="spytfyre" uid="166957" visible="true"
    timestamp="2009-09-04T13:35:42Z">
    <tag k="name" v="The Blue Blazer"/>
    <tag k="amenity" v="pub"/>
  </node>
</osm>
```

When we wish to draw your attention to a particular part of a code block, the relevant lines or items are set in bold:

```
<osm version="0.6" generator="OpenStreetMap server">
  <node id="107775" lat="51.5072647" lon="-0.1278328" version="29"
    changeset="2628959" user="EdinburghGael" uid="170586"
    visible="true" timestamp="2009-09-25T23:04:28Z">
    <tag k="place" v="city"/>
    <tag k="name:zh" v="伦敦"/>
    <tag k="name:sv" v="London"/>
    <tag k="name:sk" v="Londýn"/>
    ...
    <tag k="is_in" v="England, United Kingdom, UK, Great Britain,
      Europe"/>
    <tag k="capital" v="yes"/>
    <tag k="name:fr" v="Londres"/>
    <tag k="name:cy" v="Llundain"/>
  </node>
</osm>
```

Any command-line input or output is written as follows:

```
Kosmos.Console.exe bitmapgen compton.kpr compton.png  -mb 51.20868 -
0.63794 51.22011 -0.61772 -z 16
```

New terms and **important words** are shown in bold. Words that you see on the screen, in menus or dialog boxes for example, appear in the text like this: "You can do this by right-clicking on your **My Computer** icon and selecting **Properties**."

There are some images that are referred in the chapters but are not actually present in it. Such images can be found along with the code files on the Packt website. The images are referred as follows: "The road's name is **Down Lane**, which can be seen in the image `121120009090.jpg`."

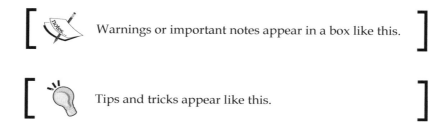

Warnings or important notes appear in a box like this.

Tips and tricks appear like this.

Reader feedback

Feedback from our readers is always welcome. Let us know what you think about this book—what you liked or may have disliked. Reader feedback is important for us to develop titles that you really get the most out of.

To send us general feedback, simply send an e-mail to feedback@packtpub.com, and mention the book title via the subject of your message.

If there is a book that you need and would like to see us publish, please send us a note in the **SUGGEST A TITLE** form on www.packtpub.com or e-mail suggest@packtpub.com.

If there is a topic that you have expertise in and you are interested in either writing or contributing to a book, see our author guide on www.packtpub.com/authors.

Customer support

Now that you are the proud owner of a Packt book, we have a number of things to help you to get the most from your purchase.

Downloading the example code for this book

You can download the example code files for all Packt books you have purchased from your account at http://www.PacktPub.com. If you purchased this book elsewhere, you can visit http://www.PacktPub.com/support and register to have the files e-mailed directly to you.

Errata

Although we have taken every care to ensure the accuracy of our content, mistakes do happen. If you find a mistake in one of our books—maybe a mistake in the text or the code—we would be grateful if you would report this to us. By doing so, you can save other readers from frustration and help us improve subsequent versions of this book. If you find any errata, please report them by visiting http://www.packtpub.com/support, selecting your book, clicking on the **errata submission form** link, and entering the details of your errata. Once your errata are verified, your submission will be accepted and the errata will be uploaded on our website, or added to any list of existing errata, under the Errata section of that title. Any existing errata can be viewed by selecting your title from http://www.packtpub.com/support.

Piracy

Piracy of copyright material on the Internet is an ongoing problem across all media. At Packt, we take the protection of our copyright and licenses very seriously. If you come across any illegal copies of our works, in any form, on the Internet, please provide us with the location address or website name immediately so that we can pursue a remedy.

Please contact us at copyright@packtpub.com with a link to the suspected pirated material.

We appreciate your help in protecting our authors, and our ability to bring you valuable content.

Questions

You can contact us at questions@packtpub.com if you are having a problem with any aspect of the book, and we will do our best to address it.

1
Making a Free, Editable Map of the World

Maps have always been a source of power. Armies have used them to give themselves a military advantage, traders have used them to find the shortest route between places of supply and markets, and fictional pirates have used them to find buried treasure. Those possessing a map have an advantage over those that don't.

It's no wonder then that maps and the information used to create them have been fiercely protected by companies and governments that own them. Many government and commercial mapping agencies charge high fees to use their data, impose strict restrictions on what you can do with the information, and may even make you pay to use maps you've drawn yourself. The OpenStreetMap project aims to change this by giving everyone their own map, for free, and to use for whatever they like.

In this book, you will learn:

- What the OpenStreetMap project is
- Why it's different from other sources of free maps
- Why you should contribute to the project
- How to add and edit geographic data
- How to turn that data into maps you can use for anything you like

This chapter explains how and why the project was started, how freely redistributable geographic data is different from maps that are merely free of charge to use, and some of the things the project has already achieved.

What is OpenStreetMap?

OpenStreetMap (http://www.openstreetmap.org/) is a project to build a free geographic database of the world. Its aim is to eventually have a record of every single geographic feature on the planet. While this started with mapping streets, it has already gone far beyond that to include footpaths, buildings, waterways, pipelines, woodland, beaches, postboxes, and even individual trees. Along with physical geography, the project also includes administrative boundaries, details of land use, bus routes, and other abstract ideas that aren't apparent from the landscape itself.

The database is built by contributors, usually called mappers within OpenStreetMap, who gather information by driving, cycling, or walking along streets and paths, and around areas recording their every move using **Global Positioning System (GPS)** receivers. This information is then used to create a set of points and lines that can be turned into maps or used for navigation. The next image shows the raw GPS recordings of courier vans working in Central London, where you can see how the many streets in the city are laid out.

Most mappers are volunteers working on the project in their spare time, although both commercial organizations and government bodies have started to contribute to the project.

The process of using groups of people to work on a task in this way, called **crowdsourcing**, is a recent phenomenon based around using the Internet to distribute tasks and gather the results. It's used by voluntary projects and commercial organizations alike, and has been particularly effective since broadband Internet connections became widely available in the western world.

Other data is gathered from out-of-copyright maps, public domain databases (ones with no copyright protection), or in some cases donations of proprietary databases by the companies that own them. In most cases, this needs further work to update and tidy the data, but it allows mappers to cover areas they can't get to, or for features that are difficult to survey on foot.

The database uses a wiki-like system where any mapper can add or edit any feature in any area, and a full editing history is kept for every object. This means any mistakes or deliberate vandalism can be rolled back, keeping the data accurate. OpenStreetMap doesn't use an existing geographic information system (GIS) to store its data, but instead uses its own software and data model to make the crowdsourcing process as easy as possible, and to allow the maximum level of flexibility in what gets mapped and how.

While the primary aim of the project is to collect geographic data, members of the project have also produced a wide range of software (much of it open source) that creates, edits, manipulates, or uses the data in some way. We'll use a selection of this software in the examples in this book.

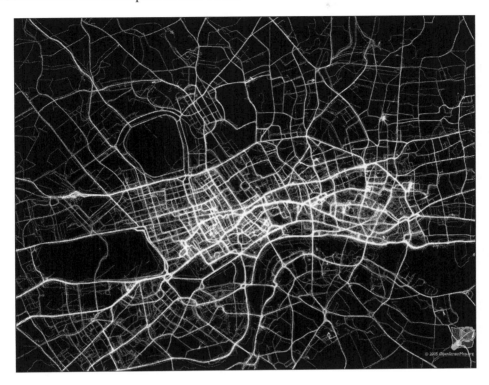

OpenStreetMap's data is free to use by anyone, for any purpose. It is released under a license that allows you to copy, change, and redistribute the data. There are many maps available on the web, all of which are free to use, and some can be embedded in your own web pages or used in mash-ups, but they also have restrictions on what you can do with those services and the data they provide.

In contrast to OpenStreetMap, none of these free services allow you to modify or redistribute their data. If the data is wrong, you can submit a bug report and hope that they fix it, but update cycles are typically slow, and can take months, if not years, to make the corrections. If the service is withdrawn, you're left with nothing. If you want to use the data offline or in an alternative format, you can't. Some even go so far as to claim ownership of any information you display on top of their maps. To borrow a phrase from open source software, these services are "free as in beer, not free as in speech".

OpenStreetMap is often compared to Wikipedia, and there are many similarities between the projects. They both create freely licensed content. They both use the Internet to allow contributors from all over the world to participate, and they both rely on collaborative editing to improve the information they contain incrementally. They both rely on a large and diverse community to ensure the project runs smoothly, and to prevent errors, whether accidental or intentional, from lowering the quality of their information.

Unlike Wikipedia, OpenStreetMap is building a database of information rather than a series of articles. The biggest difference between the projects is that to contribute to OpenStreetMap, you will, at some point, have to leave your computer behind and go out and gather some data.

Despite the name, OpenStreetMap isn't just about maps, and certainly not just about one or two maps; routing, geocoding (finding coordinates for a given object), and spatial analysis are other applications for the data. Even the maps on the OpenStreetMap website are just examples, and you can make your own maps in whatever style you choose.

Why use OpenStreetMap?

Unlike most sources of geographic data, there are virtually no restrictions on what you can do with OpenStreetMap data. You can use it for any purpose, including commercial activities, without having to pay license fees. You can edit the information in any way you like and publish the results. You can give the data to someone else without needing permission, and they can, in turn, pass it on. Your only obligation is to "share alike"; that is, to allow anyone you give OpenStreetMap data to redistribute it themselves. The particular license used is currently the Creative Commons Attribution Share-Alike 2.0 license, usually abbreviated as CC-BY-SA, although a process to change the license to the Open Database Licence 1.0 (ODBL) is currently being considered.

OpenStreetMap contains details of much more than just roads. All mappers are free to add any geographic object they find, so people have added phone boxes, bus stops, parks, public toilets, places of worship, and much more. Footpaths and cycleways are better represented in OpenStreetMap than many other databases, and there are already a number of cycling map and routing projects based on OpenStreetMap data.

OpenStreetMap is updated more often than any other geographic database. In fact, it's continually updated, and the latest data is always available to download. While proprietary databases may be updated frequently, those changes often aren't released to customers very quickly, leaving users with an out-of-date database for some time, possibly months. In contrast, the full OpenStreetMap database, known as the **Planet file**, is released every week, with updates ("diffs") released on a daily, hourly, and minute-by-minute schedule.

You can correct errors in OpenStreetMap data yourself, and share those corrections with everyone. In the UK, the problem of heavy goods vehicles using satellite navigation and being routed along narrow country lanes appears in the news regularly, and one man even found himself driving to the edge of a cliff by following his SatNav. Anyone encountering mistakes in the information like this can change the map themselves directly, and the changes show up in the database immediately.

Most of the other mapping providers only have maps in a limited number of styles, and obtaining custom maps can be an expensive and lengthy process. In contrast, with OpenStreetMap, the only limits to the number of mapping styles you can have is your own technical and cartographic ability. Several map rendering packages and Geographic Information Systems support OpenStreetMap data, and all the software you need to create maps is available free of charge..

Why contribute to OpenStreetMap?

You don't have to map anything yourself to make use of OpenStreetMap data, so why should you do any mapping yourself?

It's fun! Mappers tell stories of finding footpaths less than a mile from their house or workplace that they didn't know existed. Surveying has taken mappers to places they'd otherwise never have visited, and made them appreciate their local area in a different way.

The area you need may not have been mapped in OpenStreetMap yet. While urban areas in Western Europe have extensive coverage, more rural areas in those countries or places elsewhere in the world may still not have been mapped. The project mostly relies on volunteers, and less densely populated areas get mapped at a slower rate. Surveying the area yourself, at whatever level of detail you require, is easy to do. If you have a commercial project that requires mapping, you could pay a professional surveyor to do the work, which may be more cost-effective in the longer term than paying license fees for a proprietary database.

You may be interested in features that no one else has mapped yet. An area may have its roads extensively mapped, but not have any post boxes, car parking, or retail outlets mapped. If you add those features, you may find other mappers start adding them in other areas, and that the features you add to the database are kept up-to-date by the community. You could get back more from the project than you put in this way.

By contributing to the project, you're also contributing to the common good. The license used by OpenStreetMap means that any data in the database forms a commons, in perpetuity, and without restrictions. It puts accurate, customizable mapping within the reach of small, non-commercial groups, such as clubs and societies, charities and other voluntary organizations.

History

The OpenStreetMap project began in August 2004 when British programmer Steve Coast wanted to experiment with an USB GPS receiver he'd bought and his Linux-powered notebook. He used a piece of software called GPSDrive, which took maps from Microsoft MapPoint, breaking the license conditions. Not wanting to violate copyright on those maps, he looked around for an alternative. Coast found that there were no sources of mapping data available that he could incorporate into open source software without breaking the licensing conditions or paying huge amounts.

Coast realized that he could draw his own map, and so could others, and the project was born. After presenting his ideas at open source events in London, he found that others had similar ideas, but most hadn't got their projects off the ground. Once they were persuaded to join in, OpenStreetMap was up and running.

The idea of amateur cartographers making maps with consumer-grade equipment was greeted with some skepticism at first. Some said that standard GPS receivers were too inaccurate to make maps with, as a 10-meter error would mean roads would be in the wrong place. Others claimed that a complex infrastructure was needed for such a large-scale project. Other objectors said a predefined ontology was needed, or that the database would simply be too big.

The data model was initially very crude, consisting of simple lines drawn over Landsat information from NASA. Over a period of time, the data evolved into a more useful model, but the basic principle of imposing as few restrictions as possible on mappers was followed. The server software was initially written in Java, then rewritten in Ruby, and finally in Ruby on Rails, which is still the current server platform.

In March 2006, the first desktop editing application for OpenStreetMap, JOSM, was released. This was written in Java and allowed offline mapping for the first time. Soon after, the first full-color map was created using a renderer written specifically for OpenStreetMap, called Osmarender, showing Weybridge in Surrey, and added to Wikipedia.

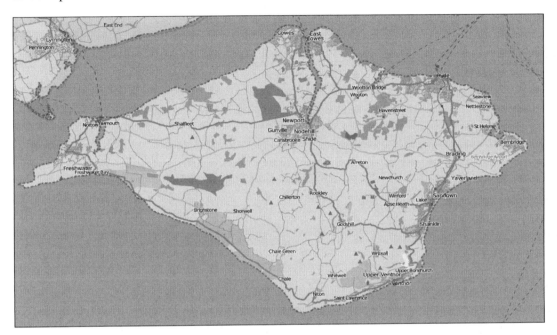

In the month of May of that year, the first mapping event was held on the Isle of Wight (pictured in the previous image). This was the first time many of the mappers had met in person, and marked a turning point for the project. One single area was mapped in detail by many people, showing that the crowdsourcing approach to geography worked. Mapping parties like this event have become regular features of the OpenStreetMap community, and are held all over the world wherever mappers spot an area that needs better coverage.

The **OpenStreetMap Foundation** was formed in August 2006 to own the infrastructure needed to run the project and accept donations. Prior to this, the servers, domain names, and other infrastructure had been owned by Steve Coast, and establishing the foundation gave the project an existence beyond one person's involvement.

May 2007 saw the server software move to the Ruby on Rails platform, and the release of the online Flash-based editor, Potlatch. Later that month, the first annual OpenStreetMap conference, **The State of the Map**, was held in Manchester. By August, there were five million ways in the database, and 10,000 registered users on the OpenStreetMap website.

In September 2007, Automotive Navigation Data (AND), a Dutch mapping company, donated its dataset for the Netherlands to the project. This was the first time a commercial organization had provided data to the project, and the first time the whole of any country was covered for some types of data.

During the same month, the process of importing the US government public domain geodata set, known as TIGER (Topologically Integrated Geographic Encoding and Referencing) was started, completing in January 2008. Work began immediately to clean this data up and to make it conform with OpenStreetMap data standards.

By February 2008, there were 25,000 registered users of the OpenStreetMap website, and an appeal was launched to raise £10,000 for new servers. In the end, over £15,000 was raised in less than a week.

State of the Map 2008 was held in Limerick, Ireland in July. By the end of that year, over 300 million points had been plotted in the database. On March 17, 2009, the 100,000th user registered on the site, less than five years after the project started; the number of users had quadrupled in just over a year.

At the time of writing, over half a million points were being added to the OpenStreetMap database each day, and the rate of growth was still increasing.

Does it work?

The process of crowdsourcing geographic data is accurate. A 2008 study by Muki Haklay at University College London examined the OpenStreetMap data for the UK in general, and London in particular, and compared it to similar data from Ordnance Survey. Haklay found that where data existed in OpenStreetMap, 80% of it coincided with Ordnance Survey data for the same features. The average distance from an OpenStreetMap feature to an Ordnance Survey equivalent was six metres, less than the general accuracy of GPS positioning, and this didn't take into account the possibility of the Ordnance Survey data being inaccurate.

There were also large areas where there was no data in OpenStreetMap. The conclusion is that if there is a problem with OpenStreetMap data, it is one of completeness rather than positional accuracy. Everything that's there is in the right place, but not everything is mapped yet.

Another vote of confidence in the quality of OpenStreetMap's data comes from one UK local authority, Surrey Heath Borough Council, whose staff have mapped the area covered by the council, and is using OpenStreetMap maps on its website and internally. You will also find maps based on OpenStreetMap data on `whitehouse.gov`, the website of the President of the United States.

Structure of the project

There is very little formal structure to the OpenStreetMap project, and what structure there is reflects the voluntary nature of contributions to the project. There are no super-users who can change the data in ways that "normal" users can't, and there are no fixed ontologies for the data. The data and the project are both intended to give mappers as much flexibility as possible, so they can map features as accurately as they can. While there are system administrators who look after the project's infrastructure, they don't get any more say about what gets put in the database than any other contributor.

The OpenStreetMap Foundation—the non-profit body set up to support the project—doesn't regulate where mappers work, what features they map, or how they describe them. The foundation does have a few working groups that look after various aspects of the project, but they are concerned with long-term development rather than day-to-day operation. The foundation will only intervene when the whole project is being put at risk by the actions of individual mappers.

This relative anarchy is both a strength and weakness of the project. On the positive side, it allows mappers to get on with recording information without needing permission, approval, or moderation. It allows the data to evolve over time to correct mistakes and to take account of changes without the overhead of a top-down design. It allows novel uses of the data without having to check license conditions. However, it also means that problems are addressed according to how interesting they are to individual mappers, rather than how important they may be to the project as a whole, as the volunteers aren't under any obligation to follow a plan.

The project's main website is `www.openstreetmap.org`, which is aimed at collecting and maintaining the data. Various other websites and resources are used to coordinate the project, including mailing lists, a wiki, a code repository, and an issue tracker. Other tools are also provided by individual mappers or companies, and these have their own websites.

The most important thing to remember is that the project has no paid employees. Even mappers whose job involves working with OpenStreetMap are paid by their companies, who have their own interests, and aren't there to solve other people's problems necessarily. While the OpenStreetMap community will be happy to help you solve any problems you have, you will be expected to make an effort yourself.

The OpenStreetMap Foundation

The OpenStreetMap Foundation (OSMF) is a non-profit company based in the UK that supports the project. The foundation exists to own the project's infrastructure, accept donations, organize the annual conference, and provide arbitration in the event of disputes. It was established in 2006 as a Company Limited by Guarantee—a type of company in the UK that is suited to non-profit-making organizations. This type of company doesn't have shareholders; rather its members or guarantors are expected to pay a small amount (currently £5) in the event of the company becoming insolvent. The foundation's Memorandum of Association forbids it from paying dividends or otherwise distributing any profits.

Anyone may join the OSMF, although it's not necessary to use OpenStreetMap or contribute towards it. Neither does the OSMF dictate what can or can't be mapped, or how mappers work. Although it owns the project's infrastructure, the foundation doesn't own the data stored on them unless mappers have explicitly chosen to reassign the copyright on their contributions to the foundation.

The OSMF is governed by a board of directors, elected by the membership, which oversees the general running of the foundation. There are also a number of working groups made up of members of the foundation in addition to board members, which provide recommendations about particular aspects of the project, such as licensing, promotion of the project, the annual conference, and a data working group, responsible for ensuring the accuracy of the data as a whole. The data working group will intervene if unresolvable content disputes occur, or for cases of persistent vandalism, but, in general, most issues are solved by the OpenStreetMap community.

The foundation also administers the GPStogo,scheme, which provides GPS receivers to mappers in developing countries. The scheme is aimed at improving OpenStreetMap's data in countries where the equipment needed is too expensive to buy by lending receivers to mappers who have already made some contribution to the project.

Achievements

In five years, OpenStreetMap has gone from a small project run by a few enthusiasts in London to being a global resource with thousands of users and the start of an industry based around the data it collects. It's no longer just a hobby for a few, but a serious project attracting the attention of companies and governments.

Many specialist maps have been created, including OpenCycleMap (http://www.opencyclemap.org/) showing cycle networks and routes, a Piste map for skiers and snowboarders, a hiking site combining maps with photos for guidance, and a nautical chart.

When the Beijing 2008 Olympic Games started, the photo-sharing website Flickr (`http://www.flickr.com/`) began using OpenStreetMap mapping of the city for its users to plot the location of their photos, and has continued to do so.

OpenStreetMap has also made an impact in humanitarian aid. The project has produced mapping of the Gaza Strip in the middle-east (seen in the previous screenshot) through tracing aerial images and information from aid workers in the region. The dangers faced by people working in the region and the rate at which the landscape changes has made up-to-date mapping difficult to obtain. It's hoped that OpenStreetMap's way of producing maps will change this.

Summary

The OpenStreetMap project, its data, and its community can provide maps and mapping data from a simple static image to a custom-rendered dynamic map of the world, to anyone, for any purpose, without restrictions. The data is accurate, detailed in places, and continually improving. The only limit to what you can do with the data is your own ability, and in this book we'll show you how to create, edit, and view the data, and how to turn it into a map showing the information you want, in the style you want.

2
Getting started at openstreetmap.org

OpenStreetMap is a diverse project with hundreds of thousands of people contributing data and making use of it in different ways. As a result, many of the resources that mappers have created and use are scattered around the Internet, but the project data and much of the documentation is hosted at openstreetmap.org, on servers operated by the OpenStreetMap Foundation.

As a crowdsourced project, OpenStreetMap is heavily reliant on having an active community participate in the project, and there are probably as many tools and websites aimed at allowing mappers to communicate and collaborate as there are for mapping and using the data. Mappers have created many different ways of sharing information, based on personal preference and the kind of information involved.

In this chapter, we'll cover:

- The main features of openstreetmap.org—the main website for the project
- Creating and personalizing an account on openstreetmap.org
- Using the map viewer and the tools associated with it
- Using the documentation in the OpenStreetMap wiki
- Communicating with the OpenStreetMap community using mailing lists, forums, IRC, and other channels

Not all the tools and features on the site are obvious from the front page, so we'll go on a tour of the site, and cover some other tools hosted by the project. By the end of the chapter, you should have a good idea about where to find answers to the questions you have about OpenStreetMap.

A quick tour of the front page

The project's main "shop front" is www.openstreetmap.org. It's the first impression most people get of what OpenStreetMap does, and is designed to be easy to use, rather than show as much information as possible. In the following diagram, you can see the layout of the front page. We'll be referring to many of the features on the front page throughout the book, so let's have a look at what's there:

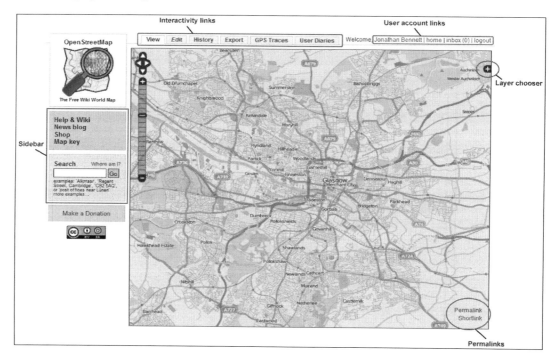

Most of the page is taken up by the map viewer, which is nicknamed the slippy map by mappers. This has its own controls, which we'll cover later in the chapter. Along the top of the map are the navigation tabs, showing most of the data management tools on openstreetmap.org. To the right of these are the user account links.

Down the left-hand side of the page is the sidebar, containing links to the wiki, news blog, merchandise page, and map key. The wiki is covered later in this and other chapters. The news blog is www.opengeodata.org, and it's an aggregation of many OSM-related blogs.

The Shop page is a page on the wiki listing various pieces of OpenStreetMap-related merchandise from several sources. Most merchandise generates income for the OpenStreetMap Foundation or a local group.

Clicking on the map key will show the key on the left-hand side of the map. As you'd expect, the key shows what the symbols and shading on the map mean. The key is dynamic, and will change with zoom level and which base layer you're looking at. Not all base layers are supported by the dynamic map key at present.

Below this is the search box. The site search uses two separate engines:

- **Nominatim**: This is an OpenStreetMap search engine or geocoder. This uses the OpenStreetMap database to find features by name, including settlements, streets, and points of interest. Nominatim is usually fast and accurate, but can only find places that have been mapped in OpenStreetMap.

- **Geonames**: This is an external location service that has greater coverage than OpenStreetMap at present, but can sometimes be inaccurate. Geonames contains settlement names and postcodes, but few other features.

Clicking on a result from either search engine will center the map on that result and mark it with an arrow.

Creating your account

To register, go to `http://www.openstreetmap.org/`, and choose **sign up** in the top right-hand corner. This will take you to the following registration form:

Create a User Account

Fill in the form and we will send you a quick email to activate your account.

Email Address:

Confirm Email Address:

Not displayed publicly (see privacy policy)

Display Name:

Your publicly displayed username. You can change this later in the preferences.

Password:

Confirm Password:

Continue

At present, you only really need an account on openstreetmap.org if you're planning to contribute mapping data to the project. Outside the main site and API, only the forums and issue tracker use the same username and password as openstreetmap.org. You don't need to register to download data, export maps, or subscribe to the mailing lists. Conversely, even if you're not planning to do any mapping, there are still good reasons to register at the site, such as the ability to contact and be contacted by other mappers.

OpenStreetMap doesn't allow truly anonymous editing of data. The OSM community decided to disallow this in 2007, so that any contributors could be contacted if necessary. If you're worried about privacy, you can register using a pseudonym, and this will be the only identifying information used for your account. Registering with openstreetmap.org requires a valid e-mail address, but this is never disclosed to any other user under any circumstance, unless you choose to do so.

It is possible to change your display name after registration, and this changes it for all current OpenStreetMap data. However, it won't change in any archived data, such as old planet files.

Once you've completed the registration form, you'll receive an e-mail asking you to confirm the registration. Your account won't be active until you click on the link in this e-mail. Once you've activated your account, you can change your settings, as follows:

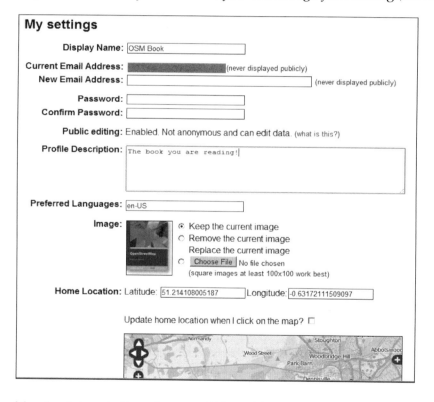

You can add a short description of yourself if you like, and add a photo of yourself or some other avatar. You can also set your home location by clicking on it in the small slippy map on your settings page. This allows other mappers nearby to see who else is contributing in their area, and allows you to see them. You don't have to use your house or office as your home location; any place that gives a good idea of where you'll be mapping is enough. Adding a location may lead to you being invited to OpenStreetMap-related events in your area, such as mapping parties or social events. If you do add a location, you get a **home** link in your user navigation on the home page that will take the map view back to that place. You'll also see a map on your user page showing other nearby mappers limited to the nearest 10 users within 50km.

If you know other mappers personally, you can indicate this by adding them as your friend on openstreetmap.org. This is just a convenience to you, and your friends aren't publicly shown on your user page, although anyone you add as a friend will receive an e-mail telling them you've done it.

Once you've completed the account settings, you can view your user page (shown in the following screenshot). You can do this at any time by clicking on your display name in the top right-hand corner. This shows the information about yourself that you've just entered, links to your diary and to add a new diary entry, a list of your edits to OpenStreetMap, your GPS traces, and to your settings. These will be useful once you've done some mapping, and when you need to refer to others' activities on the site.

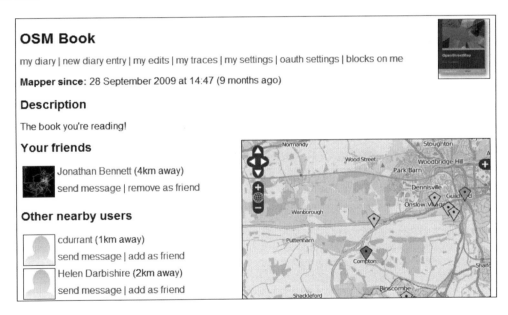

Every user on openstreetmap.org has a diary that they can use to keep the community informed of what they've been up to. Each diary entry can have a location attached, so you can see where people have been mapping. There's an RSS feed for each diary, and a combined feed for all diary entries. You can find any mapper's diary using the link on their user page, and you can comment on other mappers' diary entries, and they'll get an e-mail notification when you do.

How to use the slippy map

The slippy map is the name used to mean the interactive sliding map viewer used on the front page of openstreetmap.org. It uses an open source JavaScript library called OpenLayers, and sets of map tiles that are assembled to form a continuous, movable image. Tiles for different zoom levels show different amounts of detail, depending on the rendering rules used. Apart from being the usual way of showing the map, it's also a tool for mappers to examine the data behind the map.

You can move the map around using the controls in the top left-hand corner of the map, and zoom in and out using the slider. You can also drag the map around with your mouse or pointing device, or zoom in and out using a mouse wheel, if you have one. You can zoom in on a particular area by holding down the *Shift* key and dragging a rectangle over the area you want to see. Double-clicking on the map will re-center the view and zoom in one level.

On the right-hand side of the map, you'll see a plus sign against a blue background; this is the layer chooser. Click this to expand the box, and you'll see the list of available base layers and overlays. A base layer is the underlying map shown, while an overlay adds detail to whichever base layer you've chosen, and may include an interactive element. You can use any overlay with any base layer, but you'll only be able to interact with the top overlay.

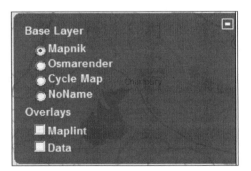

The first two base layers are named after the programs used to render them — Mapnik and Osmarender. We'll cover how both of these programs work later in the book. These layers are designed to show the amount of detail that OpenStreetMap's data contains, but may not include everything in the database, either for the sake of clarity, or because a class of feature may not be in widespread use yet. In general, you should expect to find common features, such as roads, paths, buildings, waterways, points of interest, and landuse shading on these maps. These layers are continually updated, so changes to the database usually appear within a few minutes, depending on how many mappers are making changes and how many people are viewing the map at the time.

The Cycle map is an example of a specialist rendering of the data, and shows cycle networks overlaid on a map with contour lines. It was one of the first alternative renderings of OpenStreetMap data, and has its own site at `http://www.opencyclemap.org/`. The Cycle map layer is updated less often than the first two layers, generally once a week.

The final layer, called **NoNames**, is a maintenance tool for mappers, highlighting any streets in the database without a name, where one would normally be expected. The NoNames layer was created to help mappers see where streets had been added to the database by tracing aerial imagery, which doesn't provide the names of the streets. This is only a guide, and not every road highlighted in this layer will actually have a name. It's updated once a week, so any names you add to the data won't show up straight away.

There are two overlays on the slippy map; the **maplint** overlay and the **data** overlay. Maplint shows parts of the map that use undocumented tags or has suspected errors, while the data overlay allows you to see the actual data behind a feature on the map, including any tags attached, when it was last edited and by whom. We'll cover how to use both of these overlays in Chapter 5.

At the bottom right-hand corner of the map, there are two hyperlinks, labeled **Permalink** and **Shortlink**. Both of these give you a link to your current map view, so you can bookmark it or share it with others. The permalink includes the full co-ordinates of the location, while the shortlink uses a code, and is more compact but less readable. To copy either the permalink or the shortlink, right-click on the link and choose **Copy Link Location** or your browser's equivalent.

Interacting with the data

Along the top of the slippy map are the navigation tabs for interacting with the data, or getting more information. We'll cover what most of these do in more detail in later chapters, but following a quick summary:

- The **View** tab will take you back to the map viewer, and doubles as a permalink for the current view. Clicking on this tab will reload the current view and make an entry in your browser's history. If you want to use your browser's back and forward buttons to switch between locations, you'll need to do this for each place you want to move between.

- The **Edit** tab switches to Potlatch—the online editor for OpenStreetMap. We'll cover this in Chapter 5.

- The **History** tab shows a list of recent edits in the area you're looking at. This and other tools that show what edits have taken place are discussed in Chapter 5.

> The **Edit** and **History** tabs will appear in gray if you're looking at too large an area for those tools to cope with.

- The **Export** tab takes you to the map exporter, where you can get images or data for the current area. We'll cover how to use the exporter in Chapter 6.

- The **GPS traces** tab takes you to the list of raw GPS tracks that have been uploaded to OpenStreetMap. This is where you'll add your own traces once you've collected them. Creating, uploading, and using GPS traces are explained in Chapter 3.

- The **User Diaries** tab takes you to a feed of diary entries for all OpenStreetMap contributors. At present, all diary entries are listed, rather than being filtered by location or language. This may change as more capabilities are added to the site.

Project documentation: the wiki

OpenStreetMap's data is stored in a custom-built system, but all the documentation for the project is stored in a separate wiki, using MediaWiki—the software used to run Wikipedia. It's hosted at `http://wiki.openstreetmap.org/`. In general, any information that isn't part of the map data itself is kept here.

The wiki hasn't been integrated into the login system for openstreetmap.org, so if you want to edit the wiki, you'll need to create a separate account. This isn't necessary if you only want to read the wiki.

You can use the wiki to record any information that you think might be useful to other mappers. This includes definitions of what particular tags mean in OpenStreetMap, details of how well-mapped cities, towns and villages are, instructions for installing and using various software packages, and reviews of GPS receivers.

The wiki should usually be your first resort when looking for the solution to a problem. The easiest way to find information is by using the search box. There's also a customized Google search of the wiki, mailing lists, and forum at `http://bit.ly/osmsearch`.

The most important and heavily used page on the wiki is **Map Features** (`http://wiki.openstreetmap.org/wiki/Map_Features`), for which you'll find a link in the left-hand side navigation. This contains a list of commonly used tags and descriptions of how they should be used. More unusual tags are also documented in the wiki, but you may need to use the search box to find these.

There are no restrictions on what anyone can write in the wiki, and few mappers keep it up-to-date, so some information you find in there may be out-of-date, or may not reflect standard practice by the majority of mappers. If you find a page in the wiki that looks inaccurate, you're welcome to correct it yourself, or ask in another communication channel about its contents. The wiki documents what mappers do, rather than dictating it, so you can choose to ignore its advice if it would prevent you from mapping some area or features properly.

Communicating with other mappers

As OpenStreetMap is a community-run project, you will need to communicate and collaborate with other mappers to get the most out of participating in the project. There are a number of ways you can do this, depending on your personal preferences and the type of information you're looking for (or want to offer).

Mailing lists

The main method of communication between mappers is the mailing lists. There are a large number of mailing lists reflecting the diversity of the OpenStreetMap community and the activities they're interested in. The mailing lists are hosted at `http://lists.openstreetmap.org/`. You need to be subscribed to any OpenStreetMap mailing list to post a message, and you can subscribe to any list via the web interface, or by sending an e-mail to <listname>-subscribe@ openstreetmap.org. There are five main lists that interest most mappers:

- **Announce** (`http://lists.openstreetmap.org/listinfo/announce`): Announce is a very low-volume, moderated mailing list where announcements of significant changes to OpenStreetMap are announced. You should subscribe to this list if you aren't planning to use any of the other mailing lists.

- **Talk** (`http://lists.openstreetmap.org/listinfo/talk`): Talk is for general discussion of OpenStreetMap, and is a very high-volume list. It's where most issues affecting OSM are discussed, and the discussions can be lengthy and heated sometimes. If you want to make a significant announcement, or get the attention of a large part of the OpenStreetMap community, use this list.

- **Newbies** (http://lists.openstreetmap.org/listinfo/newbies): Newbies is a mailing list for newcomers to OpenStreetMap, where you can ask simple questions. It's a lower-volume list than talk, so you won't get overwhelmed with mail by subscribing.

- **Dev** (http://lists.openstreetmap.org/listinfo/dev): Dev is where programmers writing the software that OSM runs on, and those writing the software that uses OpenStreetMap data hang out. You can use this list to discuss bugs you've found in the website or any software, but not for questions on mapping and tagging. Some OpenStreetMap software packages have their own development mailing list.

- **Legal-talk** (http://lists.openstreetmap.org/listinfo/legal-talk): Legal-talk is where licensing and copyright issues affecting the project are discussed. If you have data from an existing source you'd like to import to OpenStreetMap, this is where you should ask if the data is compatible with OpenStreetMap's license.

There are also local mailing lists for many countries, and for specific regions of some countries. You should consider joining your local list, as many country-specific tagging schemes exist, and these lists have a lower level of traffic than the main talk list.

Some specialist subject mailing lists also exist, covering particular aspects of using OpenStreetMap data, such as routing, accessibility, and geocoding. If you would like another specialist mailing list (and are prepared to operate it), you can request its creation by e-mailing the mailing lists administrator. See the mailing lists page on the wiki (http://wiki.openstreetmap.org/wiki/Mailing_lists) to find out who currently does this job.

Chatting on IRC

The best place to get live help from other mappers is on **Internet Relay Chat (IRC)**. While there's no "official" OpenStreetMap IRC server run by the project, the most popular channel is at irc://irc.oftc.net/osm, and there are many country-specific IRC channels also hosted on irc.oftc.net.

 If you don't have an IRC client installed, you can still use the main channel via the web interface at http://irc.openstreetmap.org/.

See the Contact page on the wiki (`http://wiki.openstreetmap.org/wiki/Contact`) for a list of channels. Just like the mailing lists, the main IRC channel can get very busy and the discussion heated, but any sensible question will normally get a sensible answer. The main channel is busiest and most useful during European working hours.

Forums

There is a web forum for OpenStreetMap at `http://forum.openstreetmap.org/`, where you can get in touch with other mappers. The forum isn't as heavily used as the mailing lists or IRC channels, so here you may have to wait longer to get an answer to your questions.

planet.openstreetmap.org

If you need OpenStreetMap data in bulk, you will need to download a dump of the database, known as a **planet file**. All planet files are hosted on `http://planet.openstreetmap.org/`, including the weekly full dump, the daily, hourly, and minute-by-minute change, or diff files.

There's no user interface to this site, only the raw directory listings. The latest weekly planet file is always at `http://planet.openstreetmap.org/planet-latest.osm.bz2`, and in each subdirectory containing diff files, there is a text file, `timestamp.txt`, which contains the timestamp of the most recent diff.

If you don't need data for the entire planet, you can trim down the planet file to a particular geographical area, and some sites provide pre-built extracts, covering particular countries or continents. We'll cover this in Chapter 8.

Reporting problems with OpenStreetMap software

OpenStreetMap's developers use an issue tracking system called Trac, hosted at `http://trac.openstreetmap.org/`, to manage bugs in the core OpenStreetMap software, such as the API and websites. You can report the bugs you find here, but only for software. Any problems with the map data itself should be reported elsewhere, and we'll cover some ways of doing that in Chapter 5. Trac uses the same username and password as the website.

OpenStreetMap on social networks

Unsurprisingly, many mappers and users of OpenStreetMap also use various social networking sites, and have created a presence for OSM there. Social networking sites are particularly useful when organizing OpenStreetMap-related events.

- Facebook group: `http://bit.ly/osmfbgroup`
- Facebook fan page for OpenStreetMap: `http://bit.ly/osmfanpage`
- LinkedIn group: `http://bit.ly/osmlinkedin`
- OpenStreetMap group on Upcoming: `http://bit.ly/osmupcoming`

Don't be afraid to ask

OpenStreetMap is built on community. If you have a problem, are unsure about how or whether you should map something, or need help using the maps in some way, use one of the methods we've covered in this chapter to ask someone. Experienced mappers would rather help you than have inadvertent mistakes in the data, or worse, have you give up mapping altogether.

Summary

OpenStreetMap is a collaborative project, and uses many different resources to allow mappers to communicate with each other. We've looked at most of these in this chapter, including:

- The main openstreetmap.org website
- The OpenStreetMap wiki
- Mailing lists
- IRC channels
- The web forum
- The issue tracker

Whenever you encounter a problem with creating, editing, or using OpenStreetMap data, you should always ask other mappers for guidance, as there may already be an answer, or the problem may be affecting more mappers. Choose which channel suits your personal preferences.

3
Gathering Data using GPS

In this chapter, we'll look at the tools and techniques used by the OpenStreetMap community to gather data using GPS, and upload it to the website, including:

- What the Global Positioning System is, and how it works
- How to set up your GPS receiver for surveying
- How to get the best signal, and more accurate positioning
- How to tell a good GPS trace from a bad one
- Ways of ensuring your survey is comprehensive
- Other ways of recording information while surveying

We'll also look at a couple of ways of gathering information without needing a GPS receiver.

OpenStreetMap is made possible by two technological advances: Relatively affordable, accurate GPS receivers, and broadband Internet access. Without either of these, the job of building an accurate map from scratch using crowdsourcing would be so difficult that it almost certainly wouldn't work.

Much of OpenStreetMap's data is based on traces gathered by volunteer mappers, either while they're going about their daily lives, or on special mapping journeys. This is the best way to collect the source data for a freely redistributable map, as each contributor is able to give their permission for their data to be used in this way.

The traces gathered by mappers are used to show where features are, but they're not usually turned directly into a map. Instead, they're used as a backdrop in an editing program, and the map data is drawn by hand on top of the traces. This means you don't have to worry about getting a perfect trace every time you go mapping, or about sticking exactly to paths or roads. Errors are canceled out over time by multiple traces of the same features.

OpenStreetMap uses other sources of data than mappers' GPS traces, but they each have their own problems: Out-of-copyright maps are out-of-date, and may be less accurate than modern surveying methods. Aerial imagery needs processing before you can trace it, and it doesn't tell you details such as street names. Eventually, someone has to visit locations in person to verify what exists in a particular place, what it's called, and other details that you can't discern from an aerial photograph.

If you already own a GPS and are comfortable using it to record traces, you can skip the first section of this chapter and go straight to **Techniques**. If you want very detailed information about surveying using GPS, you can read the *American Society of Civil Engineers* book on the subject, part of which is available on Google Books at `http://bit.ly/gpssurveying`. Some of the details are out-of-date, but the general principles still hold.

If you are already familiar with the general surveying techniques, and are comfortable producing information in GPX format, you can skip most of this chapter and head straight for the section *Adding your traces to OpenStreetMap*.

What is GPS?

GPS stands for **Global Positioning System**, and in most cases this refers to a system run by the US Department of Defense, properly called NAVSTAR. The generic term for such a system is a Global Navigation Satellite System (GNSS), of which NAVSTAR is currently the only fully operational system. Other equivalent systems are in development by the European Union (Galileo), Russian Federation (GLONASS), and the People's Republic of China (Compass). OpenStreetMap isn't tied to any one GNSS system, and will be able to make use of the others as they become available. The principles of operation of all these systems are essentially the same, so we'll describe how NAVSTAR works at present.

NAVSTAR consists of three elements: the space segment, the control segment, and the user segment.

- The **space segment** is the constellation of satellites orbiting the Earth. The design of NAVSTAR is for 24 satellites, of which 21 are active and three are on standby. However, there are currently 31 satellites in use, as replacements have been launched without taking old satellites out of commission. Each satellite has a highly accurate atomic clock on board, and all clocks in all satellites are kept synchronized. Each satellite transmits a signal containing the time and its own position in the sky.

- The **control segment** is a number of ground stations, including a master control station in Colorado Springs. These stations monitor the signal from the satellites and transmit any necessary corrections back to them. The corrections are necessary because the satellites themselves can stray from their predicted paths.

- The **user segment** is your GPS receiver. This receives signals from multiple satellites, and uses the information they contain to calculate your position. Your receiver doesn't transmit any information, and the satellites don't know where you are. The receiver has its own clock, which needs to be synchronized with those in the space segment to perform its calculations. This isn't the case when you first turn it on, and is one of the reasons why it can take time to get a fix.

Your GPS receiver calculates your position by receiving messages from a number of satellites, and comparing the time included in each message to its own clock. This allows it to calculate your approximate distance from each satellite, and from that, your position on the Earth. If it uses three satellites, it can calculate your position in two dimensions, giving you your latitude (lat) and longitude (long). With signals from four satellites, it can give you a 3D fix, adding altitude to lat and long. The more satellites your receiver can "see", the more accurate the calculated position will be. Some receivers are able to use signals from up to 12 satellites at once, assuming the view of the satellites isn't blocked by buildings, trees, or people. You're obviously very unlikely to get a GPS fix indoors.

Many GPS receivers can calculate the amount of error in your position due to the configuration of satellites you're using. Called the **Dilution of Precision (DOP)**, the number produced gives you an idea of how good a fix you have given the satellites you can get a signal from, and where they are in the sky. The higher the DOP, the less accurate your calculated position is. The precision of a GPS fix improves with the distance between the satellites you're using. If they're close together, such as mostly directly overhead, the DOP will be high. Use signals from satellites spread evenly across the sky, and your position will be more accurate. Which satellites your receiver uses isn't something you can control, but more modern GPS chipsets will automatically try to use the best configuration of satellites available, rather than just those with the strongest signals. DOP only takes into account errors caused by satellite geometry, not other sources of error, so a low DOP isn't a guarantee of absolute accuracy.

The system includes the capability to introduce intentional errors into the signal, so that only limited accuracy positioning is available to non-military users. This capability, called Selective Availability (SA) was in use until 1990, when President Clinton ordered it to be disabled. Future NAVSTAR satellites will not have SA capabilities, so the disablement is effectively permanent. The error introduced by SA reduced the horizontal accuracy of a civilian receiver, typically to 10m, but the error could be as high as 100m. Had SA still been in place, it's unlikely that OpenStreetMap would have been as successful.

NAVSTAR uses a coordinate system known as **WGS84**, which defines a spheroid representing the Earth, and a fixed line of longitude or **datum** from which other longitudes are measured. This datum is very close to, but not exactly the same as the Prime Meridian at Greenwich in South East London. The equator of the spheroid is used as the datum for latitude. Other coordinate systems exist, and you should note that no printed maps use WGS84, but instead use a slightly different system that makes maps of a given area easier to use. Examples of other coordinate systems include the OSGB36 system used by British national grid references. When you create a map from raw geographic data, the latitudes and longitudes are converted to the x and y coordinates of a flat plane using an algorithm called a projection. You've probably heard of the Mercator projection, but there are many others, each of which is suitable for different areas and purposes.

What's a GPS trace?

A GPS trace or tracklog is simply a record of position over time. It shows where you traveled while you were recording the trace. This information is gathered using a GPS receiver that calculates your position and stores it every so many seconds, depending on how you have configured your receiver.

If you record a trace while you're walking along a path, what you get is a trace that shows you where that path is in the world. Plot these points on a graph, and you have the start of a map. Walk along any adjoining paths and plot these on the same graph, and you have something you can use to navigate. If many people generate overlapping traces, eventually you have a fully mapped area. This is the general principle of crowdsourcing geographic data. You can see the result of many combined traces in the following image. This is the junction of the M4 and M25 motorways, to the west of London. The motorways themselves and the slip roads joining them are clearly visible.

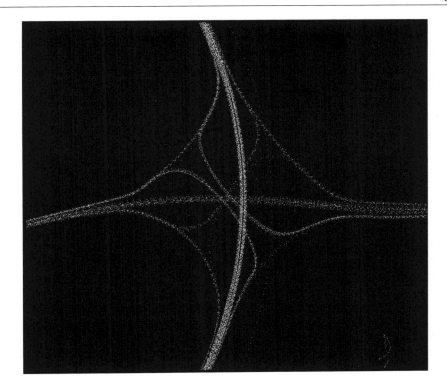

Traces are used in OpenStreetMap to show where geographical features are, but usually only as a source for drawing over, not directly. They're also regarded as evidence that a mapper has actually visited the area in question, and not just copied the details from another copyrighted map. Most raw GPS traces aren't suitable to be made directly into maps, because they contain too many points for a given feature, will drift relative to a feature's true position, and you'll also take an occasional detour.

Although consumer-grade GPS receivers are less accurate than those used by professional surveyors, if enough traces of the same road or path are gathered, the average of these traces will be very close to the feature's true position. OpenStreetMap allows mappers to make corrections to the data over time as more accurate information becomes available.

In addition to your movements, most GPS receivers allow you to record specific named points, often called waypoints. These are useful for recording the location of point features, such as post boxes, bus stops, and other amenities. We'll cover ways of using waypoints later in the chapter.

What equipment do I need?

To collect traces suitable for use in OpenStreetMap, you'll need some kind of GPS receiver that's capable of recording a log of locations over time, known as a track log, trace, or breadcrumb trail. This could be a hand-held GPS receiver, a bicycle-mounted unit, a combination of a GPS receiver and a smartphone, or in some cases a vehicle satellite navigation system. There are also some dedicated GPS logger units, which don't provide any navigation function, but merely record a track log for later processing. You'll also need some way of getting the recorded traces off your receiver and onto your PC. This could be a USB or serial cable, a removable memory card, or possibly a Bluetooth connection. There are reviews of GPS units by mappers in the OpenStreetMap wiki.

There are also GPS receivers designed specifically for surveying, which have very sensitive antennas and link directly into geographic information systems (GIS). These tend to be very expensive and less portable than consumer-grade receivers. However, they're capable of producing positioning information accurate to a few centimeters rather than meters.

You also need a computer connected to the Internet. A broadband connection is best, as once you start submitting data to OpenStreetMap, you will probably end up downloading lots of map tiles. It is possible to gather traces and create mapping data while disconnected from the Internet, but you will need to upload your data and see the results at some point. OpenStreetMap data itself is usually represented in Extensible Markup Language (XML) format, and can be compressed into small files. The computer itself can be almost any kind, as long as it has a web browser, and can run one of the editors, which Windows, Mac OS X, and Linux all can.

You'll probably need some other kit while mapping to record additional information about the features you're mapping. Along with recording the position of each feature you map, you'll need to note things such as street names, route numbers, types of shops, and any other information you think is relevant. While this information won't be included in the traces you upload on openstreetmap.org, you'll need it later on when you're editing the map. Remember that you can't look up any details you miss on another map without breaking copyright, so it's important to gather all the information you need to describe a feature yourself.

- A paper notebook and pencil is the most obvious way of recording the extra information. They are inexpensive and simple to use, and have no batteries to run out. However, it's difficult to use on a bike, and impossible if you're driving, so using this approach can slow down mapping.

- A voice recorder is more expensive, but easier to use while still moving. Record a waypoint on your GPS receiver, and then describe what that waypoint represents in a voice recording. If you have a digital voice recorder, you can download the notes onto your PC to make them easier to use, and JOSM—the Java desktop editing application—has a support for audio mapping built-in.

- A digital camera is useful for capturing street names and other details, such as the layout of junctions. Some recent cameras have their own built-in GPS, and others can support an external receiver, and will add the latitude, longitude, and possibly altitude, often known as geotags, to your pictures automatically. For those that don't, you can still use the timestamp on the photo to match it to a location in your GPS traces. We'll cover this later in the chapter.

Some mappers have experimented with video recordings while mapping, but the results haven't been encouraging so far. Some of the problems with video mapping are:

- It's difficult to read street signs on zoomed-out video images, and zooming in on signs is impractical.

- If you're recording while driving or riding a bike, the camera can only point in one direction at once, while the details you want to record may be in a different direction.

- It's difficult to index recordings when using consumer video cameras, so you need to play the recording back in real time to extract the information, a slow process.

Automatic processing of video recordings taken with multiple cameras would make the process easier, but this is currently beyond what volunteer mappers are able to afford.

Smartphones can combine several of these functions, and some include their own GPS receiver. For those that don't, or where the internal GPS isn't very good, you can use an external Bluetooth GPS module. Several applications have been developed that make the process of gathering traces and other information on a smartphone easier. Look on the Smartphones page on the OpenStreetMap wiki at http://wiki.openstreetmap.org/wiki/Smartphones.

Making your first trace

Before you set off on a long surveying trip, you should familiarize yourself with the methods involved in gathering data for OpenStreetMap. This includes the basic operation of your GPS receiver, and the accompanying note-taking.

Configuring your GPS receiver

The first thing to make sure is that your GPS is using the WGS84 coordinate system. Many receivers also include a local coordinate system in their settings to make them easier to use with printed maps. So check in your settings which system you're getting your location in. OpenStreetMap only uses WGS84, so if you record your traces in the wrong system, you could end up placing features tens or even hundreds of meters away from their true location.

Next, you should set the recording frequency as high as it will go. You need your GPS to record as much detail as possible, so setting it to record your location as often as possible will make your traces better. Some receivers can record a point once per second; if yours doesn't, it's not a problem, but use the highest setting (shortest interval) possible. Some receivers also have a "smart" mode that only records points where you've changed direction significantly, which is fine for navigation, but not for turning into a map. If your GPS has this, you'll need to disable it. One further setting on some GPSs is to only record a point every so many metres, irrespective of how much time has elapsed. Turning this on can be useful if you're on foot and taking it easy, but otherwise keep it turned off.

Another setting to check, particularly if you're using a vehicle satellite navigation system, is "snap to streets" or a similar name. When your receiver has this setting on, your position will always be shown as being on a street or a path in its database, even if your true position is some way off. This causes two problems for OpenStreetMap: if you travel down a road that isn't in your receiver's database, its position won't be recorded, and the data you do collect is effectively derived from the database, which not only breaks copyright, but also reproduces any errors in that database.

Next, you need to know how to start and stop recording. Some receivers can record constantly while they're turned on, but many will need you to start and stop the process. Smartphone-based recorder software will definitely require starting and stopping. If you're using a smartphone with an external Bluetooth GPS module, you may also need to pair the devices and configure the receiver in your software.

Once you're happy with your settings, you can have a trial run. Make a journey you have to make anyway, or take a short trip to the shops and back (or some other reasonably close landmark if you don't live near shops). It's important that you're familiar with your test area, as you'll use your local knowledge to see how accurate your results are.

Checking the quality of your traces

When you return, get the trace you've recorded off your receiver, and take a look at it on your PC using an OpenStreetMap editor (see Chapter 5) or by uploading the trace. Now, look at the quality of the trace. Some things to look out for are, as follows:

- Are lines you'd expect to be straight actually straight, or do they have curves or deviations in them? A good trace reflects the shape of the area you surveyed, even if the positioning isn't 100% accurate.

- If you went a particular way twice during your trip, how well do the two parts of the trace correspond? Ideally, they should be parallel and within a few meters from each other.

- When you change direction, does the trace reflect that change straight away, or does your recorded path continue in the same direction and gradually turn to your new heading?

- If you've recorded any waypoints, how close are they to the trace? They should ideally be directly on top of the trace, but certainly no more than a few meters away.

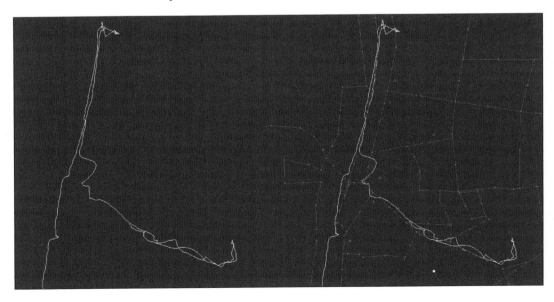

The previous image shows a low-quality GPS trace. If you look at the raw trace on the left, you can see a few straight lines and differences in traces of the same area. The right-hand side shows the trace with the actual map data for the area, showing how they differ.

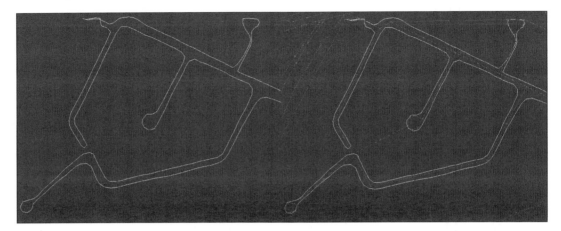

In this image, we see a high-quality GPS trace. This trace was taken by walking along each side of the road where possible. Note that the traces are straight and parallel, reflecting the road layout. The quality of the traces makes correctly turning them into data much easier.

If you notice these problems in your test trace, you may need to alter where you keep your GPS while you're mapping. Sometimes, inaccuracy is a result of the make-up of the area you're trying to map, and nothing will change that, short of using a more sensitive GPS. For the situations where that's not the case, the following are some tips on improving accuracy.

Making your traces more accurate

You can dramatically improve the accuracy of your traces by putting your GPS where it can get a good signal. Remember that it needs to have a good signal all the time, so even if you seem to get a good signal while you're looking at your receiver, it could drop in strength when you put it away.

- If you're walking, the best position is in the top pocket of a rucksack, or attached to the shoulder strap. Having your GPS in a pocket on your lower body will seriously reduce the accuracy of your traces, as your body will block at least half of the sky.

- If you're cycling, a handlebar mount for your GPS will give it a good view of the sky, while still making it easy to add waypoints. A rucksack is another option.

- In a vehicle, it's more difficult to place your GPS where it will be able to see most of the sky. External roof-mounted GPS antennas are available, but they're not cheap and involve drilling a hole in the roof of your car. The best location is as far forward on your dashboard as possible, but be aware some modern car windscreens contain metal, and may block GPS signals. In this case, you may be able to use the rear parcel shelf, or a side window providing you can secure your GPS.

- Don't start moving until you have a good fix. Although most GPS receivers can get a fix while you're moving, it will take longer and may be less accurate. More recent receivers have a "warm start" feature where they can get a fix much faster by caching positioning data from satellites.

You also need to avoid bias in your traces. This can occur when you tend to use one side of a road more than the other, either because of the route you normally take, or because there is only a pavement on one side of the road. The result of this is that the traces you collect will be off-center of the road's true position by a few meters. This won't matter at first, and will be less of a problem in less densely-featured areas, but in high-density residential areas, this could end up distorting the map slightly.

Surveying techniques

You can gather traces while going about your normal business, or you can make trips specifically to do surveying for OpenStreetMap. The amount of detail you'll be able to capture on a normal journey will be far lower than during a survey, but there are still some techniques you can use to record as much detail about your surroundings as possible.

The first technique to consider is your mode of transport while mapping. For some types of features, there is only one choice: For a motorway you need to use a vehicle, and for narrow footpaths you'll need to walk.

For everything in between, you need to use some judgment. Many existing mappers have found that for suburban and residential areas, a bicycle is the most efficient way of mapping. It's faster than walking, and cheaper than a car. A bike is also easier to turn around when you reach a dead end, and you can dismount and walk along paths where cycling isn't allowed.

Making your survey comprehensive

To make sure you map an area completely, you need to be methodical about how you travel around the area. One simple rule that works well in suburban and residential areas is to "always turn left". This is a standard technique used by robots to find the layout of a maze (and thus escape from it), and it works just as well for mapping.

What "always turn left" means is that if you come across a left-hand turn you haven't previously been down, then take it. Unless the streets you're mapping have a grid pattern, you'll eventually come to a dead end. When you do, turn around and head back down the street, and start turning left again. This method isn't perfect, particularly when there are loops in the road network, so take notes of places where you pass turnings on the opposite side of the road and make sure you visit them later in your survey.

If you're mapping on a bike or in a car in a country where traffic drives on the right, then "always turn right" works better, but the choice is yours. For streets in a grid layout, a simple back-and-forth sweep of the grid should be fine.

When mapping trunk roads or motorways with grade-separated junctions, remember to map the on-and off-ramps whenever you can. While the roads themselves get mapped quite thoroughly by people on long-distance journeys, individual junctions can get missed out even if the roads they join to are mapped. If you make a regular journey along a road like this, why not try to map one junction per day, which shouldn't take much of your time, but will still increase the map coverage quite quickly.

What level of detail you map streets at is up to you, but some typical features you could gather include bridges, changes of speed limit, route reference numbers, and street names.

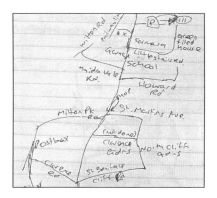

Along with the streets, there will be point features and areas to map. Point features include street furniture, such as postboxes, bus stops, and public telephones; while areas can be things, such as car parks, playgrounds, and pedestrianized areas. You can record the relative locations of features that you find in a notebook, like in the previous diagram. Your diagram doesn't need to be neat and precise, as your GPS receiver is recording the location of each feature you find. All your notes have to do is record the information, such as names, that don't automatically get stored in your GPS traces.

You can mark the location of point features using your GPS's waypoint feature. Simply stop right by the object you want to mark and press the waypoint button, or choose the option from the menu. You can either use the automatic name given to the waypoint and add extra notes to your notebook, or use a voice recorder. Alternatively, rename the point with a more descriptive name.

For areas, you have a choice of techniques. You can either walk around the perimeter of the area, and the trace will show a loop giving the feature's location. Alternatively, you can just mark the corners of the area as you pass them using waypoints, and note which waypoints are linked to make the area. The latter method is useful when the area is large and you'll be passing by its corners anyway as part of your survey. As with all other OpenStreetMap data, your first pass at mapping an area doesn't have to be perfect, so don't be afraid of drawing an approximation based on three corners of an area.

Photo mapping

You can use a digital camera to take pictures of road signs, street names, or junction layouts to speed up your mapping and help increase the accuracy of your mapping.

The first step to take is to synchronize the clock on your digital camera with the time on your GPS receiver. This is because we'll be using the timestamp information on each photo to match it up with its location, based on your GPS trace. Not all digital cameras allow you to set the time to the nearest second, but as long as the difference between the clocks isn't too great, this won't cause a problem. One way of coping with a difference between the clocks is to take a picture of your GPS receiver with the time showing. You can use this to work out the offset between your camera and the GPS clock.

Once you've done this, you can use your camera to record lots of details you'd otherwise have to note by hand, and because you can place photos by their datestamp, you don't even need to record waypoints on your GPS.

Take pictures of more things than you think you need, and in more detail than you think. For instance, street name signs may also contain a part or all of the street's postal code, or the district it's in; take your picture from too far away, and you won't be able to read it. The previous photo captures two pieces of information: the name of the road, and the speed limit to the left of the sign. We can also see another road to the left, which gives us another clue which way the camera was facing when the picture was taken.

Some built-up areas may be very difficult to get an accurate GPS fix, so taking photos can give you an idea of where any straight sections are, and where the road bends, which may not be obvious from a poor GPS trace.

You can photograph road junctions to record their layout. It can be difficult and time consuming to record a trace of every possible path through a junction, so a series of pictures can help you map the junction accurately in less time.

Once you have your pictures and the accompanying GPS trace, you can load them into an OpenStreetMap editing application and use the information to draw the map. We'll cover how to do this in Chapter 5.

OpenStreetMap doesn't have any image hosting facilities itself, so if you want to make your pictures available for other mappers to use, you'll need to use a separate photo-hosting site, such as Flickr. There is a sister project called OpenStreetPhoto, but this is still under development and at present only provides detailed mapping for the Benelux countries.

Audio mapping

Like photo mapping, audio mapping can speed up the surveying process, but it has an additional advantage of being useful if you're just making a regular journey. The principle is that you describe features into a voice recorder. It's possible to do audio mapping with a tape-based voice recorder, but for best results, a digital recorder that timestamps the files it creates, is used. A smartphone may have a similar function built-in, or you may be able to add some software to it that will allow you to take voice notes. If you do use a tape recorder, remember to record a waypoint in your GPS, then start your voice note with the waypoint name.

As with photo mapping, you'll need to synchronize the clock on your recorder with that on your GPS. Some difference is OK, but if you're mapping high-speed roads, you need to keep it to a minimum; at 60mph or 100kph, you're travelling 22 meters every second. At that rate, a 20 second difference between the clocks would put your mapping out by over 500 meters.

Once you've done that, just use your voice recorder to describe your surroundings. You could record street names by saying, "Turning left into High Street" just before a turn, or use the past tense just after the turn. If a street name has an unusual spelling, or could be spelled in several different ways, it's a good idea to spell out the name in your voice note. It's a good idea to note route names or numbers occasionally as well, even if you've already noted them, as it removes some uncertainty.

Getting your traces into the right format

Once you've completed your mapping, you need to get the information off your receiver and into the right format. The simplest ways of getting traces off your GPS are using a direct cable connection or a removable memory card. Some recent receivers can act as USB mass storage devices, allowing you to get your traces onto your PC with a simple file copy. For older units, you may have to use the software that came with your GPS.

OpenStreetMap only accepts traces in GPS Exchange format (GPX)—an XML vocabulary for traces and waypoints. You can find out more about the GPX vocabulary at http://www.topografix.com/gpx.asp. OpenStreetMap also only accepts GPX files with timestamps on each trackpoint in the trace. This is to prevent mappers from uploading traces that have been converted from an existing map database, which will usually be subject to copyright and will therefore, not be suitable for use in OpenStreetMap. This doesn't present a problem most of the time, as practically all receivers store the time in traces made in real time.

If your receiver doesn't produce GPX files itself, you'll need to translate from its own format into GPX. Fortunately, there's an application that can convert most formats into GPX, called GPSBabel. If you can get GPX files from your receiver, you can skip this section.

GPSBabel is a free software package, and you can download it from http://www.gpsbabel.org/ for Windows, Mac OS X, and Linux. While GPSBabel itself is a command-line application, graphical interfaces are available for Windows, Mac, and Linux. The Windows version is shown in the following screenshot, which shows the command line it creates for the options you select in the interface.

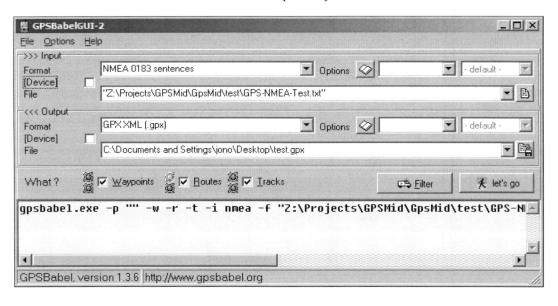

We'll work through an example of converting a file in NMEA format to GPX using GPSBabel on Windows, but the procedure is the same for any file format. GPSBabel can also communicate directly with many models of GPS, and this can speed up the conversion process over downloading and converting separately.

GPSBabel is a powerful package with many options, and you won't need most of them. We're interested in converting tracks and waypoints into the GPX format. For an NMEA file, you'd use the following command line:

```
gpsbabel -w -t -i nmea -f <input filename> -o gpx -F <output filename>
```

If you're using the Windows command line, you'll need to use `gpsbabel.exe` as the program name.

The first two options tell GPSBabel to process waypoints and tracks; the default is to only process waypoints, which is of no use to OpenStreetMap. The third option is the input file format to use. You specify this with the `-i` flag, a format identifier, and the filename. The list of every identifier GPSBabel understands is available in the online documentation or by using the program's help option:

```
gpsbabel -h
```

You set the output format using the `-o` option in a similar manner, and in our case, this is always -gpx. The input filename is specified in the `-f` option, and the output file in the `-F` option. The input can also be a device if you want to retrieve traces directly from your GPS. Check the GPSBabel documentation for the precise syntax needed for your device.

As already mentioned, OpenStreetMap only accepts traces with valid timestamps. If you want to conceal the time you made a journey, you can do so through a filter that will change the start time of your trace. As timestamp information is useful to other mappers and for future projects, if you're going to change your timestamps, you're asked to do so to an obviously fake time so that your trace can be filtered out from automatic processing.

Use the following option to adjust timestamps:

```
gpsbabel -w -t -i nmea -f <input filename> -x track,start=19700101000000
-o gpx -F "<output filename>
```

This will set the start of your trace to January 1, 1970, well before OpenStreetMap was started, so it will be obvious that these are faked timestamps.

If you want to disguise the precise location your traces start or stop at, you can use a filter to discard any points within a given radius of a point. The following command line will filter out points within 500 meters of 10 Downing Street—the British Prime Minister's residence:

```
gpsbabel-w -t -i nmea -f <input filename> -x radius,exclude,distance=0.5K
,lat=51.5034052,lon=-0.1274766 -o gpx -F <output filename>
```

You can use more complex filters to clean up your traces by taking actions such as discarding any points with a high DOP. You don't need to do this, as the crowdsourcing process eliminates such errors over time, but it can help when first mapping an area.

Adding your traces to OpenStreetMap

Once you have your traces in GPX format, you can upload them to openstreetmap. org. Whether you need to upload your traces depends on which editing method you prefer; we'll cover editing in Chapter 5. If you want to use the online editor, Potlatch, you need to upload your traces. If you use a desktop editor such as JOSM or Merkaator, you don't need to upload your traces to use them, but it's still useful to OpenStreetMap as a project if you do.

It's hoped that in the future, the traces can be put to other uses, including automatically generating average speeds for roads, detecting one-way streets and the layout of junctions. This automatic processing isn't yet in place, but the more the data it has to work with (once it's working), the better.

Some editors and applications support direct upload of traces, but the simplest way of adding your traces to OpenStreetMap is via the website. The upload form is at the top of the list of your traces. To find this, make sure you're logged into the site, and browse to the front page. Click on **GPS Traces** among the tabs at the top of the map, then click on **See just your traces, or upload a trace**.

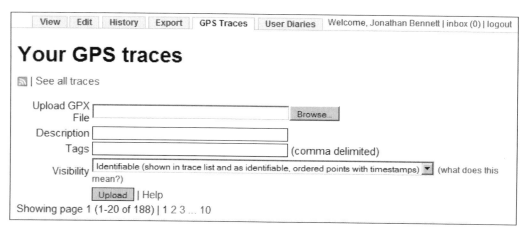

You should now see the upload form, shown in the previous screenshot. The top field is the file to be imported. Either enter the path to your GPX file directly, or use the **Browse** button to find it on your hard drive. You can only upload one trace at a time using the web interface at present.

You need to add a short description for each file you upload. You can also add tags to each trace, so you can find particular sets of traces at a later date. There are no folders or other ways of organizing your traces, so adding tags at upload is a good idea. Note that these aren't the same tags as those added to OpenStreetMap data, which we'll cover in Chapters 4 and 5. Enter a list of comma-separated tags in the form field. You can use as many tags as you find useful.

You can set the privacy level for a trace using the Visibility drop-down. All points in traces uploaded to OpenStreetMap are visible in some way, but you can choose whether other mappers can see the trace as a whole, and whether they can see who uploaded the trace. Some mappers feel that because many of their traces start or end at their home or workplace, other users shouldn't be able to see whose traces are whose, but there have been no reported incidents of a mapper's privacy being invaded through a trace they've uploaded. You can use GPSBabel to remove all points within a given radius of your start and end points, which will have a similar effect.

You have four privacy options:

- **Private**: hides all details of your traces from other users, and only shares the individual points in each trace. Other users can't see which points come from which trace, or who uploaded them. The trace isn't shown in the public list of traces.

- **Public**: is a historical option and isn't useful for newly uploaded traces. It shows the trace in the public list of traces, but still anonymizes the points when an area is downloaded. This was previously the only option other than **Private**.

- **Trackable**: allows other users to see the timestamps for each point, but not who uploaded them. The traces don't appear in the public list of traces.

- **Identifiable**: allows other users to see the entire trace, including who uploaded it. The trace is shown in the public list of traces, and the original file is available for download for everyone.

Which option you choose is entirely up to you, but your traces will be of more use to the OpenStreetMap project if you use one of the last two options.

After you've completed the upload form, you can press the **Upload** button to submit the trace. It will then go into a queue to be processed. Processing can be almost instant, or can take several minutes, depending on how many mappers are uploading traces. Once processing is complete, you'll get an e-mail to your registered address telling you whether the import was successful or not. Common reasons for imports to fail include a lack of timestamps, or GPX files that contain only waypoints, not tracks.

Refresh your traces list, and you should now see your newly imported trace, complete with a thumbnail image, and the description and tags you entered.

20090620_171623.gpx ... (3,329 points) ... 3 months ago more / map / edit PUBLIC
Part of NCR7 between Kenmore and Pitlochry
by Jonathan Bennett in Sustrans, Scotland, NCR7, Kenmore, Pitlochry

Clicking on a trace's filename or the **more** link will take you to a details page for that trace, where you can also edit the description, tags, and visibility of that trace. You can't edit a trace itself in OpenStreetMap, although JOSM does have support for GPX file editing. You won't normally need to make changes to GPS traces, but if you do, you'll need to edit a local copy of the file, delete the version on the site, and upload the new one.

The **map** link in a trace entry will take you to a map view centered around the trace's starting point. The **edit** link will take you to Potlatch—the online editor—with that trace loaded, so you can add the features it represents to the map. We'll cover this in Chapter 5.

Clicking on the uploader's name will take you to his/her user page. Click on any of the tags to see any traces you've tagged with that word. You'll also see a list of your tags in the sidebar on the left of the screen. Note that when you're looking at an individual trace page, the tag links in the page and in the sidebar will take you to a list of all traces on openstreetmap.org with that tag, not just yours.

Collecting information without a GPS

It's still possible to gather data for OpenStreetMap without a GPS receiver. Roads near your area may have been mapped by someone passing through the area, so will be missing any detailed information and points of interest. There are also areas where mappers have traced aerial images or out-of-copyright maps without surveying the area in person, so these areas will need details to be filled in.

You can identify areas that have been traced purely from aerial images using the **NoName** layer in the slippy map. This highlights any residential roads without a name, so any large blocks where all streets are highlighted are likely to have come from aerial images.

To go mapping without a GPS, you'll need a hard copy of the existing map of an area. You can print directly from the slippy map, but you'll get better results by using the Export facility to produce a PDF map of the area you're going to survey. We'll cover how to do this in Chapter 6. You can then print this out with far greater control over layout. Note that you can't export an image of the NoName layer itself, but if you have a Garmin GPS receiver, you can download an add-on map version of NoName from Cloudmade (http://downloads.cloudmade.com/).

Once you have your printed map, grab a pencil and head out. Mark the map with the locations of any points of interest, or any missing streets and paths. Don't worry about precise positioning, as it's more important to have data in an approximate location than have no data at all. You or another mapper can refine the positioning later. Use the techniques described earlier in the chapter to ensure you cover the whole of the area you've chosen.

There's a site aimed at making the process of mapping from an existing map more streamlined. Walking Papers (http://walking-papers.org/) is a site aimed at increasing participation in OpenStreetMap beyond dedicated mappers in possession of a GPS receiver. The site allows you to choose an area of the map and print it out, so that you can make notes on it. Walking Papers uses a different style of cartography than the slippy map, so there's more room around features to write and draw.

You then scan the printed map and upload it to the site. A 2D bar code on the map allows Walking Papers to match your annotated map to the area it was taken from, and use it as a background in your editor. You can then add data to the map by drawing over your scan and adding points of interest where you've marked them. Even if you don't have a scanner, you can use Walking Papers' more spacious maps as your template to draw on.

Have you finished?

Once you've surveyed an area, it's time to turn the information you've gathered into a map, and that's what the next two chapters cover. However, don't be fooled into thinking that this means the end of surveying.

The most obvious reason for needing to re-survey an area is that there have been changes on the ground, such as road layout changes, buildings being demolished and built, or the opening and closure of shops or other amenities. One of OpenStreetMap's great strengths is the ability to update it and have the new data available to everyone immediately. You'll know of construction work taking place in your local area, so you can be ready to map the new features as soon as they're finished.

You may also stumble across new roads or road layouts while making an unrelated journey; another good reason to keep your GPS receiver recording, even if you're not planning to do any surveying.

Apart from changes in the real world, your skill as a surveyor will grow the more you do it, so revisiting an area will allow you to capture more detail with greater accuracy than you may have done the first time you visited. Remember, you can record any permanent geographical feature in OpenStreetMap, and ultimately it's hoped to include anything that can be mapped.

Summary

Surveying for OpenStreetMap isn't as difficult as you might first think, and it certainly doesn't need expensive, complex equipment. We've seen that you can do surveys with one or more of:

- An inexpensive consumer-grade GPS receiver, or even no GPS at all
- A notebook and a pencil
- A digital camera
- A voice recorder

We've also learned some basic surveying techniques, including:

- Covering an area methodically
- Recording as much detail as possible, even if it's not immediately obvious how it will be used
- Using a combination of recording devices to speed up and improve the accuracy of mapping
- Surveying an area multiple times to capture changes to features, or to increase the overall accuracy of the data.

4

How OpenStreetMap Records Geographical Features

OpenStreetMap uses its own data model, which is different from any other data model used by Geographic Information Systems (GIS). It uses only three primitive types, combined with a free-form tagging scheme that allows you to describe accurately, virtually any geographic feature. It also describes the topology of the features—how they are connected, and how you move from one to another. This last feature is essential for some applications, particularly routing, but isn't a feature of many traditional GIS'.

In this chapter, we'll look at how OpenStreetMap's systems record the features you draw on the map. In particular, we'll look at:

- OpenStreetMap's design goals
- The data structures used
- The tagging system
- The guidelines used by the OpenStreetMap community to make the data as consistent and accurate as possible.

The simplicity of the data model often surprises people coming to OpenStreetMap from a traditional GIS background, who are used to many different layers of data, and having each feature described independently of others. Some look for a fixed ontology, or for more complex data structures, and when they don't find them, they are a little lost.

Although OpenStreetMap's data model and software developed over time in small jumps, and without any particular project planning, there are some principles which you can treat as the design goals for OpenStreetMap's systems:

- The data model and API are as simple as possible—"the simplest thing that will work". The data model is designed to be easy to create and edit data, rather than to be simple to render or use in other applications. You're expected to do a certain amount of post-processing on any data you take from OpenStreetMap to make it fit your application's needs, and many tools to help with this processing have already been created by the OpenStreetMap community.

- The system has wiki-like editing, where multiple users can make edits to the same area at the same time, and a full history of edits is kept. Adapting an existing GIS to fit these requirements would have been difficult, and probably more work than creating the OpenStreetMap server stack.

- The system needs to record topology—how features are connected to each other—as well as their positions, which many existing GIS don't record. Many potential uses of OpenStreetMap data aren't possible without recording the topology.

- The system should allow maximum freedom to mappers to record the features they find, with the minimum of overhead and bureaucracy.

You should note that very little data checking is done by the OpenStreetMap server. As long as the data submitted is consistent, it won't be refused. There are many third-party data-checking tools for OpenStreetMap, and we'll cover these in the next chapter, but the server itself only ensures that the basic data meets certain formatting standards, and after that allows anything.

Let's look at what the data model consists of before we get into the nitty-gritty of editing. We're still looking at things in an abstract way, but it helps to understand the model, as every editor has a slightly different way of working.

Data primitives

There are three basic primitives in the data model: **nodes**, **ways**, and **relations**. The geographical features these represent can be described by tagging them with key-value pairs. In mathematical terms, the OpenStreetMap data model is a mixed graph; it consists of corners or vertices, and edges. Different parts of the graph may be connected, or may be isolated, depending on what features in the real world they're modeling.

The default format for representing the data model is XML, and that's the only format in which you can currently retrieve data from the OpenStreetMap server. The data model can be represented in other formats, and indeed Potlatch — the Flash-based online editor in use at openstreetmap.org — uses Flash's Action Message Format to communicate with the server. We'll cover how you can retrieve individual objects from the OpenStreetMap API in Chapter 8, but you don't need to know how the API works just to edit the map.

There are several attributes common to every primitive type. Each has a numerical ID, but these are only unique within each type, so there could be a node, way, and relation all with the same ID number. .

Every version of every object in OpenStreetMap is preserved in the database, and so any object you get from the API contains a version number, along with the ID of the changeset — a set of changes made at the same time — in which that version was created. The display name and user ID of the last editor to touch the object is also returned. The reason both of these items are returned is that mappers are free to change their display name at any time, but their user ID is always the same, so it allows you to trace the creator of a particular version of an object even if they've changed their name.

 Note that the OpenStreetMap API will always return the current display name, rather than the name in use at the time of the edit, so this provision only applies when dealing with older copies of the data.

The `visible` attribute specifies whether the version of the object you have is current data. If this is set to `false`, that version of the object is old, and has been superseded by a more recent version or deleted. The API always returns the most recent version by default, but it is possible to get the full edit history of any object if you need it. There's also a timestamp for each object, showing when it was last updated, but you should always use the version number as your indication of whether or not an object has changed.

Nodes

Nodes are points in space. They are the only primitives to have position information, and all other primitives rely on nodes for their location. A node can be used on its own to map a point of interest, as a junction between two ways, or just as a change in direction of a way.

The XML for a node looks like the following:

```
<osm version="0.6" generator="OpenStreetMap server">
  <node id="483034256" lat="55.9458449" lon="-3.2035477" version="1"
    changeset="2369219" user="spytfyre" uid="166957" visible="true"
    timestamp="2009-09-04T13:35:42Z">
    <tag k="name" v="The Blue Blazer"/>
    <tag k="amenity" v="pub"/>
  </node>
</osm>
```

The XML for a feature is created by an editing application or the OpenStreetMap API, so you will never need to create this information by hand, but it does allow you to see the structure of the information. This particular node is a pub in Edinburgh, and it is rendered on a map in the following image:

Each node has its latitude and longitude stored in decimal format at up to 7 decimal places:

```
lat="55.9458449" lon="-3.2035477"
```

This gives a latitudinal resolution of 1cm, and a longitudinal resolution of around 1cm at the equator, or around 0.6cm at Greenwich. This level of accuracy is far greater than you can measure using consumer-grade GPS equipment, and on par with that achievable with professional surveying equipment. In short, it's accurate enough for any purpose you're likely to put OpenStreetMap data to.

The OpenStreetMap API doesn't check for duplicate nodes, so it's possible to have two nodes in exactly the same place. This has happened in the past where data has been imported to OpenStreetMap incorrectly, and there are tools that will check for duplicate nodes.

The only child elements of a node are the tags applied to that node. We'll cover how tagging works later in the chapter, but the XML for any tag is a single element tag with k and v attributes for the key and value respectively.

Ways

Ways are ordered list of nodes. They can describe linear features, such as roads, paths, and waterways. They can also be closed to form areas. Where they're used to describe linear features, the way should normally be placed down the center line of the physical feature, and at the perimeter for an area.

A way exists semi-independently of the nodes it contains, in that a node itself doesn't change by being a part of a way or not, and the data for a way doesn't change when its nodes are changed, including being repositioned. This means that if all the nodes that make up a way are moved, but the nodes in the way or the way's tagging are left the same, no new version of the way will be created, and no change will show up in the way's history.

A way must have at least two nodes, and can, at present, have a maximum of 2000 nodes. The upper limit is a practical measure to stop very long ways affecting the performance of openstreetmap.org's servers, rather than an inherent limit in the data model. A node can belong to more than one way, and there is no limit to how many it can belong to.

The ordering of nodes in a way is preserved, so they are always returned in an order in the original upload. Ways are considered to have direction (although this isn't always significant), depending on what the way is being used to describe. The direction is from the first to the last node, and most editors can show this direction using arrows.

If the first and last node of a way are the same, the way forms a closed area, which can be tagged to indicate what the area represents. If a more complex shape than a simple polygon is needed to describe a feature, such as a building with a courtyard in the middle, or a forest with a clearing, this is done using several ways and a relation to link them. While there's no technical restriction that prevents a way from being used as an area from self-intersecting, this will almost certainly create problems with renderers' and should be avoided.

Ways can cross without being joined, and you should only assume ways are physically connected if they share a node, or more than one node. Conversely, you should only join two ways with a node if they actually physically connect. The best example of this is a bridge carrying one road over another. It isn't possible to stop halfway along the bridge and turn onto the road below, so the data you use to represent the bridge and the road below should reflect this. We'll discuss how this is done in an editor, in the next chapter.

The OpenStreetMap server doesn't check the geometry of a way at all. It can be anything from a straight line to a scribble, as long as the nodes used are all visible. A node can even be used more than once in a way, as this is required for areas and ways that loop round onto themselves, but there's no check at the API whether nodes are reused sensibly. Rather, it's expected that editing applications ensure that they don't produce nonsensical ways in the first place.

The XML format for a way looks as follows:

```
<osm version="0.6" generator="OpenStreetMap server">
  <way id="43157302" visible="true" timestamp="2009-10-26T10:45:09Z"
    version="1" changeset="2954960" user="Ed Avis" uid="31257">
    <nd ref="540653724"/>
    <nd ref="25507043"/>
    <nd ref="107762"/>
    <nd ref="25507038"/>
    <nd ref="107759"/>
    <tag k="highway" v="primary"/>
    <tag k="lcn_ref" v="6a"/>
    <tag k="name" v="Parliament Street"/>
  </way>
</osm>
```

This way is the southern end of Parliament Street in London, shown here on the map:

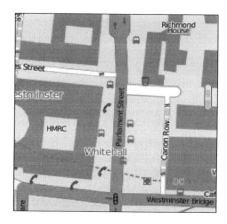

Each way has the common attributes for all primitives, but none specific to ways. The child elements of the way are the tags and the list of nodes. A different XML element, nd, is used to reference nodes, rather than to declare them to make parsing an OpenStreetMap XML file easier. The ref attribute specifies the ID number of the node.

Note that the description of a way doesn't include any details of which other ways it connects to. Instead, this is inferred from which of a way's nodes are shared with other ways. If two ways share a node, it's assumed that you can move from one to the other, subject to any restrictions described in the tags of either way, or a relation describing the restriction. It's normally easy to tell which ways are joined in one of the editors, but there are tools to detect possible errors.

Relations

Relations are lists of primitives, including other relations. Relations exist to allow mappers to model features that can't be described using a single node or way, or where two of the same type of feature overlap. Examples include complex, branching streets, long distance routes, or the turn restrictions at junctions.

The XML for a relation looks as follows:

```
<osm version="0.6" generator="OpenStreetMap server">
  <relation id="113421" visible="true" timestamp=
    "2009-11-03T10:08:27Z" version="2" changeset="3023369"
    user="Jonathan Bennett" uid="5352">
    <member type="node" ref="270186" role="via"/>
    <member type="way" ref="4418767" role="from"/>
    <member type="way" ref="4641665" role="to"/>
    <tag k="restriction" v="no_right_turn"/>
    <tag k="type" v="restriction"/>
  </relation>
</osm>
```

This particular relation represents a turn restriction, and doesn't render on either of OpenStreetMap's two main map styles. Instead, it's used by a routing algorithm to avoid making an illegal turn. There are many other types of relations, and this is perhaps the least mature part of OpenStreetMap's data model and mapping.

The relation element has the common attributes for all primitives, and no others. Child elements are the list of tags and the list of relation members. Each member element has attributes, giving its type and ID number, along with a role. The role attribute is a simple string whose values and significance are defined by the type of the relation itself. The role can be left blank for relation types that don't require one. In our example, the type is given as a restriction, and the members show the route that's illegal to take using the node representing the junction with the turn restriction, and the ways representing the roads that's illegal to access.

The members of a relation are returned in the same order they were created in, but the order isn't necessarily significant, and may not even be the desired order depending on the application that created the relation in the first place. If you have an application that uses relations in an order-sensitive way, you should check the documentation for that relation type to see whether a convention for member order has been established.

As with the other primitives, the OpenStreetMap API doesn't check whether a relation conforms to any particular format or standards, so you should do your own validation before using any relations.

Changesets

There is a further data structure used in the OpenStreetMap API—the Changeset. Changesets aren't part of the data model proper, but are simply groups of changes made to data primitives. Changesets were introduced to make it easier to identify related changes to the map, and to tidy up the attribution of data. Before their introduction, the only way of identifying edits that changed related parts of the map was to search for groups of edits by one editor within a short time of each other, which was time-consuming and prone to error. As OpenStreetMap grows in popularity, vandalism is expected to be more of a problem, and changesets are one of the tools used to keep the data accurate.

Changesets are opened by editor applications before uploading any edits, then closed once they're done. Every edit to OpenStreetMap data must be assigned to a changeset. One mapper can have any number of changesets open at any time, although idle changesets are closed automatically by the server after one hour. We'll cover the precise details of using changesets in Chapter 8, and most OpenStreetMap editing software automatically creates and closes changesets for you.

The XML for a changeset looks as follows:

```
<osm version="0.6" generator="OpenStreetMap server">
  <changeset id="1787116" user="Jonathan Bennett" uid="5352"
    created_at="2009-07-10T09:52:28Z" closed_at=
    "2009-07-10T09:52:31Z" open="false" min_lon="-4.4224237"
    min_lat="56.1382502" max_lon="-3.9613266" max_lat="56.6330162">
    <tag k="created_by" v="JOSM"/>
    <tag k="comment" v="Additions and edits to NCR7"/>
  </changeset>
</osm>
```

The first thing you'll note is that the changeset XML doesn't contain any details of the changes themselves. This information is stored in the OpenStreetMap database, and you can download a separate XML file containing the changes, but they're kept separate from the changeset description.

The changeset description XML tag contains, like other primitives, an ID, again only unique to changesets, the display name of the mapper, and his/her user ID. The XML also contains timestamps for when the changeset was opened and closed, and the four corners of the bounding box for the changeset. The bounding box describes the smallest box containing all the edits in the changeset, and can be very large if the edits are far apart.

A changeset can be tagged in the same way as other primitives, but with an important exception: Once a changeset is closed, it can't be edited. Changesets are a record of mapping activity, rather than of geodata, so they aren't editable, particularly as attribution of OpenStreetMap's data relies on the tags attached to a changeset.

Tagging

The physical aspects of OpenStreetMap's data are contained within the data model and primitives, but it's the tagging of each primitive that tells you what all the features are, and where the real power of the model lies. The data primitives have tags added to them that tell you what real-world features they represent, and from that you can render a map, produce a routing mesh, or build some other application.

A tag is a simple key-value pair. Any primitive can have an arbitrary number of tags attached to it. The keys and values are simple strings, and can contain any valid Unicode characters upto a 255 character limit. The OpenStreetMap server currently uses the UTF-8 encoding scheme.

That is the entire technical specification for tagging in OpenStreetMap. The API itself doesn't enforce any more rules on tags than the ones just mentioned. Instead, what gets put in tags is decided by the OpenStreetMap community, through custom, discussion, and consensus. There are already a huge number of keys and values in use, some on millions of objects, and some that have only been used a few times.

Tagging is semantic, in that you describe the properties of a feature you're mapping, rather than describing how you'd like it to appear on the map. The appearance of any feature on a map based on OpenStreetMap data is entirely under the control of the person who configures the renderer for that map. There's no way to force a particular feature to look a particular way in any renderer.

One analogy is with the languages web pages are written in, HTML and Cascading Style Sheets (CSS). Ideally, the HTML for a web page contains no layout information at all, but simply describes the structure of the document. A style sheet is then used to turn that document into a visual or aural representation based on the rules in the style sheet. Similarly, OpenStreetMap's data only describes the structure of a map, and leaves the presentation to a style sheet in the map renderer.

You should also remember that not all applications of OpenStreetMap data involve rendering a map. Routing, geolocation, and statistics applications may not use a map at all, or may ignore any tags that have no bearing on their algorithms.

The OpenStreetMap wiki is used to document what a particular key-value pair means. The entry is normally written by the first person to use the tag, but some tags may not be documented, or may have their page in the wiki written by someone else.

The normal way of writing down a tag is in the form

```
key=value
```

for a fully qualified tag, or

```
key=*
```

for a key with no particular value. You'll see this notation used on the wiki, mailing lists, and other documents used in OpenStreetMap. Most editors use separate text boxes for the key and value for a tag, so you won't normally need to write out a tag in full.

Keys and symbolic values are normally written in lowercase, without any whitespace, with underscores to aid readability where necessary. The OpenStreetMap API isn't in itself case sensitive, but it doesn't translate the case of any tags you add. Most tools for processing and rendering OpenStreetMap data are case sensitive, so literal values such as street names should be written with the correct capitalization and spacing. British English is normally used for keys and symbolic values, even for features in non-English speaking countries. Although this is mostly because the project was started in the UK, it's proved useful in allowing the same tags to be used in different countries.

Only one value per key is supported, so if you need to record multiple values for a particular key, the normal practice is to use a prefix or suffix to the tag, separated by a colon. The most common use of this is for place names in multiple languages. For example, the node for London gives its name in many different languages, as follows:

```
<osm version="0.6" generator="OpenStreetMap server">
  <node id="107775" lat="51.5072647" lon="-0.1278328" version="29"
    changeset="2628959" user="EdinburghGael" uid="170586"
    visible="true" timestamp="2009-09-25T23:04:28Z">
```

```
<tag k="place" v="city"/>
<tag k="name:zh" v="伦敦"/>
<tag k="name:sv" v="London"/>
<tag k="name:sk" v="Londýn"/>
<tag k="name:ru" v="Лондон"/>
<tag k="name:fi" v="Lontoo"/>
<tag k="name:es" v="Londres"/>
<tag k="name:gv" v="Lunnin"/>
<tag k="name:zh_pyt" v="Lúndūn"/>
...
<tag k="is_in" v="England, United Kingdom, UK, Great Britain,
    Europe"/>
<tag k="capital" v="yes"/>
<tag k="name:fr" v="Londres"/>
<tag k="name:cy" v="Llundain"/>
</node>
</osm>
```

In the preceding example, you can see that the name is specified by tagging with `name:<lang>=<value>`, where `lang` is the two-letter ISO language code. Other examples of this include the `addr:` prefix for tags giving addressing information for streets and buildings, and `piste:` for mapping skiing pistes and their properties.

While there are few absolute rules that have to be followed in OpenStreetMap, there are some guidelines the community has come up with to guide mappers when editing the data.

"Any tags you like"

One important principle within OpenStreetMap is that there are no restrictions on what key-value pairs can be used at all, whether technical or policy. You are free to use whatever combination you think best describes the feature you're mapping. This principle is summed up in the phrase "Any tags you like".

Taken literally, this could sound like an invitation to chaos. Instead, what it actually means is that you don't need prior approval to use a new tag, should you need to. You're still expected to use the existing keys and values, where they accurately describe the feature you're trying to map. However, you may come across a type of feature that hasn't been mapped before, and need to come up with a new tag. This isn't just possible, it's encouraged.

If you do create a new tag, you are expected to add documentation for your new tag to OpenStreetMap wiki, describing what you have used it to represent, otherwise other mappers won't know what the tag means, even if you think your tag is self-explanatory. Documentation is also translated into other languages, so the mappers doing the translations need an original text to work from.

"Don't tag for the renderer"

You'll come across this phrase regularly on the OpenStreetMap mailing lists, and it's another important point to remember. What it means is: Don't use an incorrect tag to produce a particular effect in a map renderer.

For example, you might be tempted to tag a building as woodland so that it appears on the map in green, because the building itself is painted green. This creates many problems:

- A different render of the map may use a different color for woodland, or may ignore woodland altogether, so your building won't appear as you intended, and may not appear at all.

- If a geocoding application looks for a feature tagged as a building with a particular name, it won't check your feature, because it's not listed as a building, but as an area of woodland. The converse may also happen, where someone searches for woodland, finds your building, and gets incorrect information.

If you want a map where certain features appear in particular colors or styles, you can render your own map, and we'll show you how to do this in Chapters 6 and 10.

Verifiability

The information recorded in OpenStreetMap should be **verifiable**. This means that another mapper should be able to resurvey the information you've added, and come to the same conclusion as you have, based on what they are able to observe. This applies both to the primitives you use, and the tags you attach to them. Your GPS traces are the method of verifying the position of the features you've mapped, but the tags you use should also be verifiable.

- You should only tag features based on what can be seen "on the ground". If verifying the features you're adding needs access to an additional source of information, it shouldn't go in OpenStreetMap. For instance, if you're going to add a named hiking trail to OpenStreetMap, this should be based on the signposts you find along the trail, not a guidebook.

- You should only add tags that are objective in nature. If another mapper could have a different opinion about a particular tag, it isn't a good tag. Things to avoid include relative descriptions of features ("large", "steep") where objective measurements can be used, such as "15m" or "5% gradient", or subjective judgments, such as the difficulty of cycling along a particular road.

Verifiability doesn't mean that the observations have to be possible using the naked eye, but any measurements that need specialist equipment to make are unlikely to become commonly used in the OpenStreetMap community, and you won't see other, similar features appearing outside the area you're mapping.

Above all, documenting what you mean when you use a particular tag, and checking what others have documented will help improve the verifiability of your mapping.

A few core tags

There are too many tags in use in OpenStreetMap to document them all here. There is a core set of tags that get used most often to represent commonplace features, and they'll cover the majority of the mapping you need to do, unless you have very specialist needs.

The main list of common tags is kept at `http://wiki.openstreetmap.org/wiki/Map_Features`, and the three main OpenStreetMap editing applications have built-in presets that reflect standard practice within the community. Although there's no technical distinction, there are generally two types of tags, namely, ones which define which class of feature a primitive is, and ones that define additional properties. We'll cover the class tags here, and introduce some of the property tags while we're editing in the next chapter.

Not every tag that's been documented is listed in Map Features, and not every tag that's been used is documented. If you want to find other tags, you can use the wiki's built-in search, the custom Google search at `http://bit.ly/osmsearch` which searches the wiki, mailing lists and forum, or another third-party tool.

There are a few third-party tools that collect statistics on tag usage, including Tagwatch (`http://tagwatch.stoecker.eu/`) and OSMDoc (`http://osmdoc.com/`). These tools are useful for seeing which tags are popular, but they don't contain any documentation for the tags, so you'll still need to refer to the wiki. However, they do work by analyzing the contents of the OpenStreetMap database, so they're based on what people are actually using.

The most mapped features in OpenStreetMap are roads, and in some areas of the map, they're currently the only features. All roads, footpaths, and other land-based routes use the `highway=*` key. The use of the word "highway" can be confusing in countries where this is used for a particular class of road (particularly the US), but its use in OpenStreetMap is based on British use, where it means any public road, even unsurfaced ones only suitable for farm vehicles.

The main values for `highway=*` are:

- `highway=motorway` for long-distance, high-speed roads with access restrictions.

- `highway=trunk` for other long-distance roads with less strict access restrictions.

- `highway=primary` for major general purpose roads.

- `highway=secondary` for minor general purpose roads.

- `highway=tertiary` for other through routes, possibly signposted routes.

- `highway=unclassified` for through routes not covered by any of the previous tags.

- `highway=residential` for roads in residential areas (whether urban or rural), where there is generally no through traffic.

- `highway=service` for alleys, driveways, and other roads that only provide access to a single location.

- `highway=footway` for any footpath, whether surfaced or unsurfaced, rural or urban. There are more specific versions of this tag for particular types of path, but you should start by using `highway=footway`.

Amenities are the next most-used category of features. These could be anything from litter bins to shops. Amenities are usually mapped using single nodes, but where the amenity is a building, and the outline of the building is in OpenStreetMap, that should be tagged with `amenity=*`.

Some common values for `amenity=*` are:

- `amenity=parking` for public car parks. Further tags specify whether there's a charge, how many spaces are there, and whether it's outdoors or indoors.

- `amenity=place_of_worship` for a religious building of any religion or denomination. Extra tags give the religion.

- `amenity=fuel` for a vehicle filling station. The particular fuels on sale can be specified in extra tags. Note that other fuels, such as solid fuels for heating, aren't mapped using this tag.

- `amenity=post_box` for a post box. The operator, collection times, and any reference number can also be tagged using `ref=*`.

- `amenity=cafe` for a cafe, possibly serving food.

- `amenity=pub` for a public bar serving alcoholic drinks.

- `amenity=hospital` for any hospital, with availability of emergency treatment specified in a separate tag, `emergency=*`.

There are many more values for `amenity=*` in use, so it's best to consult the wiki page for the key to find out how to tag any other amenities you're mapping.

Settlements are mapped using the `place=*` tag, the main values for which are:

- `place=country` for a country, although it's believed that all countries in the world are already mapped
- `place=county`
- `place=city` for very large settlements, although what exactly is a city varies from country to country
- `place=town`
- `place=village`
- `place=hamlet`
- `place=suburb`

Most places are currently mapped using a single node, but in the future, areas may be used where appropriate. You should always add a `name=*` tag with a `place=*` tag.

The `waterway=*` key covers all linear water features, such as rivers, as follows:

- `waterway=river`
- `waterway=stream`
- `waterway=canal`
- `waterway=ditch`

Ways tagged with `waterway=*` should point in the direction of flow where appropriate. There are also tags for features found along waterways such as lock gates, weirs, and waterfalls.

Don't worry if there seems too much information to remember here. In the next chapter, we'll look at editing OpenStreetMap using three different editing applications, all of which have tagging presets, which means you don't have to memorize the precise tags needed for every feature.

Other useful keys

As there are so many keys and values in use in OpenStreetMap, it would be impossible to describe them all here. The following are a few other keys you will encounter in areas that have already been mapped. You can read more about them in their respective wiki pages.

- `landuse=*` and `natural=*` specify what use man has put an area to, and what naturally exists there, respectively. These two tags are always used on closed ways, and ideally the areas should be contiguous, but not overlap.

- `leisure=*` for leisure facilities, such as sports centers, sporting venues, or pitches.

- `railway=*` covers most features related to railways, including the tracks themselves, stations, and level crossings. Miniature and funicular railways are also included.

Creating a new tag

If you've found a feature you need to map, and you've looked on the wiki, searched the archives, asked in the mailing lists and on the IRC channel, but still haven't found a suitable key and value for your feature, it's time to create your own. To do this, either reuse the appropriate key and use a new value, or if necessary, use a new key.

To create a new tag, you simply use it. Choosing the text to use in your tag is less important than writing documentation for it. While the tag text will record the information you're trying to add to OpenStreetMap, it won't mean much to everyone else using the map, so it's vital you document your new tag. You do this by creating a new page in the wiki. See the guidance at `http://wiki.openstreetmap.org/wiki/Tagging` for the currently preferred way of doing this. If you'd like to get other mappers' feedback on your new tag, you can send an e-mail to the tagging mailing list, requesting comments.

When using a new tag, remember:

- Tagging is semantic, so choose a tag that describes what the feature is, not how it should appear on the map.

- Use English, unless you have a good reason not to.

- Try to be precise without being too verbose.

- It's not necessary to describe a feature completely in a single tag. You can describe an Italian restaurant using `amenity=restaurant` and `cuisine=italian`, rather than `amenity=italian_restaurant`.

- Tags should be fail-safe. If an application doesn't understand your tag, it should be able to ignore it without any side effects. Add properties to a feature, rather than trying to subtract them or creating an exception.

- Document what you do in the wiki. Include photos, hyperlinks, and other materials to show what you mean by the tag you've created. This way, your tag has more chance of being used by the community and becoming an accepted convention.

- Be prepared to change your tag if, after discussion, a better tag is suggested.

If you're in any doubt, contact the OpenStreetMap community for help with creating and documenting a new tag.

Summary

We should now have covered enough of the structures and theory in OpenStreetMap for you to be confident in drawing and tagging features.

We've covered the types of data you can record, as follows:

- Nodes, representing point features
- Ways, representing linear or area features
- Relations, representing more complex features

We've also looked at the tagging system and how the community uses and manages it:

- Tags describe a feature in a verifiable, objective way
- Use editor presets and existing tags to describe features where possible
- Look in the OpenStreetMap wiki to find tags for unusual features
- You don't need prior approval to use a new tag if you need it

In the next chapter, we'll put these to practical use, along with examples of how to map using the most common editing applications.

5

OpenStreetMap's Editing Applications

After you've gathered GPS traces, taken photos, written or spoken notes, and otherwise recorded surveyed information, it's time to turn it into data in the OpenStreetMap database. The normal way of doing this is using one of the editing applications available, each of which has its own strengths:

- **Potlatch** — the web-based editor
- **JOSM** — a Java-based desktop editor
- **Merkaartor** — a desktop editing application

Although most mappers find which editor suits them best and tend to stick to it, they're not all suitable for use under all circumstances, so it's best to be familiar with more than one. Here's a quick run-down of the differences between the editors:

	Potlatch	JOSM	Merkaartor
Requirements	Web browser with Flash (or compatible) plugin	Any operating system with Java SE 1.5 or higher	Windows, Linux, or Mac OS X with Qt
Works while offline?	No	Yes	Yes
Needs installing?	No	No	Yes
Supports photo mapping?	Partially	Yes	Yes
Supports audio mapping?	No	Yes	No

There are also differences in each editor's workflow and user interface, and we'll look at these as we go along. We'll look at getting started with each individual editor, then cover the general editing techniques that apply generally.

To help you learn how to use each of these editors, and learn more general editing techniques, we're going to work through a real-life example. We're going to map the village of Compton in Surrey, England. You can download an example GPS trace and images for the village from the code bundle available on the Packt website. After you've downloaded the ZIP file, unpack it into a new directory. The GPS trace and pictures represent a survey of the village done on bicycle, and on foot where cycling isn't allowed or practical. Both the trace and the images were gathered using a Nokia 5500 smartphone, running Mobile Trail Explorer (http://code.google.com/p/mobile-trail-explorer/)—an open source trace recording application for Java ME. The trace has also been uploaded to OpenStreetMap, where you can see it at http://bit.ly/osmtrace, which redirects you to http://www.openstreetmap.org/user/Jonathan%20Bennett/traces/564418.

If you look at the animated preview of the trace in the code bundle, you can get an idea of how it was gathered. The aim of this survey is to cover all public roads, most other rights of way, and to gather any significant points of interest around the village. Private land, house names and addressing, land boundaries, and other details will be ignored.

There's a combination of named waypoints in the GPS trace and timestamped photos to record information, such as street names and the location of points of interest. The photos themselves aren't geocoded, meaning they don't have the co-ordinates they were taken at attached to the image; instead, we're going to use JOSM and Merkaartor's built-in photo mapping features to locate them. If the GPS receiver or software you're using doesn't support named waypoints, you could write the same information in a notebook against the number of the waypoint.

Although all the editors we're looking at allow you to convert a GPS trace directly to OpenStreetMap data, there are many reasons why you shouldn't normally do this:

- The GPS trace will contain inaccuracies. You can see from our sample trace that even a good trace can sometimes be inaccurate.

- Traces of areas with dead-end roads will involve some back-tracking, producing two sections of trace for one road.

- If your trace overlaps areas of existing mapping, it will produce some duplicate data.

- Traces will contain points at regular intervals along straight sections of road where they're not necessary. Having unnecessary nodes slows down data transfer and rendering.

You'll probably find that converting and then tidying a raw trace will take as much time as drawing over the trace by hand, so most of the time you'll be faster tracing manually. The only exception to this is where you are mapping a long-distance route in a previously unmapped area, but few of these remain in industrialized western countries.

Potlatch

Potlatch, named after a native American custom of giving gifts, is a Flash-based online editor for OpenStreetMap. Its biggest advantage over other editors is that it doesn't need to be installed for you to use it. You can use any computer with a browser and a Flash plugin to edit OpenStreetMap using Potlatch, and it's safe to use on public computers, provided you log out of the site afterward.

Potlatch has fewer features than the desktop editors we'll cover, but it's still able to create and edit the same data as any other editor. It has a system of presets, which makes tagging the data you create faster and prevents typing errors, but still allows you to enter free-form keys and values for tags. Potlatch does include support for photo mapping, but the workflow currently involves uploading your pictures to a separate website, so we won't cover this feature.

Launching Potlatch

You must be logged in to openstreetmap.org to use Potlatch. Once you have logged in, there are two ways to start the editor.

The most obvious way is through the **Edit** tab at the top of the main map page, which starts Potlatch and loads data for the current map view. This will only work at zoom level 13 or above, and if you're zoomed too far out, the tab will be grayed out to let you know. If you're going to start Potlatch this way, it's a good idea to zoom in as far as possible to the area you want to edit, as this will speed up the process of loading the data.

The other way of starting Potlatch, and the one we're going to use, is from the trace list. Check if you're logged in, and go to our example trace (`http://bit.ly/osmtrace`). Next to any trace, there are two links after the starting coordinates — **map** and **edit**:

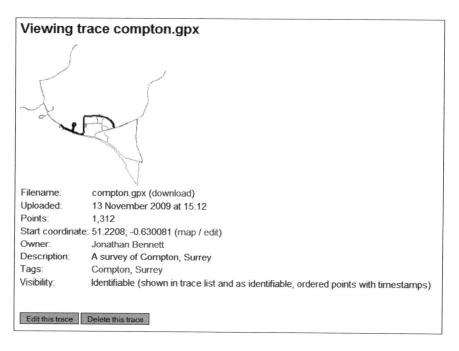

Click on **edit** and you'll be taken to the area covered by the trace, and the trace itself, along with its named waypoints, will be loaded into Potlatch. You can load the public GPS traces for any area you edit, but this won't load any waypoints from those traces.

Once you've opened Potlatch, you'll be asked whether you want to edit in a normal way, with a save button, or work "live", where your edits are written to the database as you do them. This latter way of working can be faster for simple edits, but makes it much easier to introduce errors to the data accidentally, so it should be used with caution. Live mode is most useful when two or more people are editing the same area simultaneously, such as after a mapping party, as everyone's edits will show up immediately, and there's less chance of conflict between two edits. The next version of Potlatch isn't expected to have live editing mode.

If you start Potlatch from a GPS trace, you'll also see a checkbox labeled **Convert GPS track to ways**, which you should leave unticked. Click **Edit with save**, and you should see something like the following image:

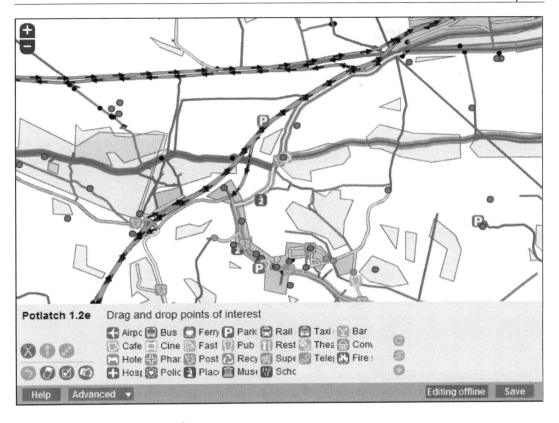

It may take some time for all the data to load, depending on the load on the OpenStreetMap servers. Potlatch will warn you with a message in the top right-hand corner of the edit pane when it's loading data. Once it has, you can start mapping.

Understanding Potlatch's user interface

Let's take a look at what we can see. Most of the Potlatch window is taken up by the map pane. In the preceding screenshot, we can see many different classes of features, shown using different colors, icons, and drawing styles:

- The green lines marked with arrows are dual carriageway (the British term for a divided highway) trunk roads, where the arrows show that the traffic is one-way. In this case, these are the A3 and A31 roads. Each carriageway is mapped using a separate way. Trunk roads may also be single carriageway, in which case only one way is used.

- A red line with a black outline is a primary road, seen at the top right in our screenshot, where the A31 turns from trunk to primary status.

- An orange-yellow line is a secondary road, and in the UK this means a B-road, while a pale yellow line is a tertiary road.

- Blue lines are water features, such as streams and rivers.

- Red lines without an outline are footpaths, and green lines without an outline are bridleways.

- Some points of interest may have an associated icon, such as parking spaces, pubs, or places of worship in our example. Others are shown using a green dot.

- Any way with a highlight is a member of a relation. The color of the highlight reflects the type of relation.

Potlatch is modeless, which means you don't have to switch tools or modes to perform different tasks. Instead, it changes action depending on what the cursor is over, and what objects, if any, are selected. You perform other actions using the buttons, dragging icons into place, or by entering the tags directly. A few commands are only available as keyboard shortcuts, and these are listed in Potlatch's online help. There's also a cheat sheet available in Scalable Vector Graphics (SVG) format at http://bit.ly/plcheatsheet.

Below the map area are Potlatch's controls. The largest section of the controls is normally taken up by the point of interest (POI) palette. In this palette is a set of icons which you can drag onto the map to add those features. This is the simplest way of adding points of interest, as it adds the data and tags in one operation.

When you select an existing feature, the palette is replaced by the tag editor. Select the main road running through Compton, and you should see something like the following screenshot:

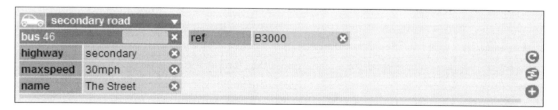

You can see the tags attached to this way, which show it's a secondary road named **The Street**, with a **30mph** speed limit, and the road number **B3000**. The dark gray box above the tags with the text **bus 46** shows that this way is a member of a relation, in this case a bus route relation. The blank box at the right of the relation is for the role, which isn't needed for a bus route. Click on the relation box, and Potlatch's relation editor will appear in the center of the window. The top box in the tag editor is the presets, which we'll explain shortly.

Over on the right-hand side are three buttons. From top to bottom, these are:

- **Repeat tags**: This button adds the tags from the last selected object of the same type as the currently selected object.
- **Add to relation**: This adds the selected object to a new or existing relation.
- **Add tag**: This adds a blank row for a new tag in the tag editor.

Once you've added a new row for a tag using this last button, you can add the tag by typing the key and value directly into the boxes. As you type, Potlatch will pop up a list of matching keys and values, which you can select using the arrow keys and the return key. If the tag you want isn't in Potlatch's list, you can continue typing.

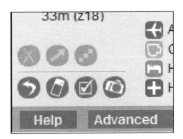

In the bottom left corner of the editor are some buttons:

- The scissors button (top left) will split a way into two at the currently selected node. In the previous screenshot, this is grayed out because nothing is selected.
- The next button showing an arrow will reverse the direction of the selected way. The arrow on the button points from the first to the last node of the way, so you can see the direction even on ways without direction arrows. Only the ways with tags explicitly stating the direction using `oneway=yes` have arrows, but all ways have a direction, and in some cases, such as waterways, the direction is implicitly significant.
- The top right button will align all the nodes in the selected way in a straight line. This tidy function has a safety feature where very curved ways won't be straightened unless you hold down the *Shift* key.
- The bottom left button is undo. You can also press Z to undo edits. Potlatch has a multi-level undo buffer, so you can undo multiple edits.
- The next button will show GPS traces for the area shown in the edit pane. Note that this button isn't a toggle, so pressing it again won't remove the traces already on screen.

- The button with a checkbox brings up the options for Potlatch. We don't need to change any of these for now.

- The camera button loads geotagged images from the source specified in the options.

At the bottom of the Potlatch window is the button bar, where you can access the online help and advanced settings, and see your current editing mode. If you're in save mode, a save button will also appear. The online help has a guide to using Potlatch, as well as a list of keyboard commands, but you can't have the guide on screen and edit at the same time.

Editing data using Potlatch

To move the map around, you click on an empty space—where there are no features drawn on the map—hold the mouse button down, and drag the map, and to zoom, you use the controls in the top left-hand corner. If you zoom or move the map so that a new area is displayed, Potlatch will automatically download the data for that area, and there may be a delay while the new area is drawn. Again, it's best to wait for the data to finish loading before entering new features. You always get the latest data from OpenStreetMap when editing in Potlatch.

To select a feature, move the cursor over it, and the cursor should change to a hand pointer. Click on the feature, and the icon palette will be replaced by the tag editor, showing the tags for the selected feature. The feature itself will also be highlighted, and in the case of a way, its nodes will be highlighted as well. Any node that's shared with other ways will have a black border around it. You can only select one feature at a time in Potlatch, so if you need to change tags on large numbers of objects at the same time, you should use one of the desktop editor applications.

Click anywhere in the empty space, and a node is placed, and Potlatch starts drawing. If you click on the same node again, you stop drawing, but the node is still selected, allowing you to add tags to it. If you want to add points of interest, it's often quicker to use the icons at the bottom of the edit pane. You can drag these into place, and Potlatch adds a new node with the correct tags for that type of feature to the map.

To draw a new way, click to place the starting node, then draw the way node by node, clicking again on the last node to stop drawing. You can start a new way from an existing node by holding down *Shift* and clicking on the node. You then start drawing as normal. To end a way on an existing node, simply draw the way and click on the node last, then click on it again to stop drawing.

You can stop drawing at any time by pressing *Return*. If you select a node at the end of a way, Potlatch will automatically start drawing, so if you just want to edit the tags on that node, press *Return* to stop drawing and leave the node selected. If you want to remove just the last node you placed, press *Delete* or *Backspace*. You can also press *Escape* to stop drawing and delete any unfinished way you've started to draw. This can be useful if you've tried to move the map, but accidentally started drawing instead.

To delete a single node in Potlatch, select it, and press *Delete* or *Backspace*. To delete an entire way, select it, and press *Shift* and *Delete*.

Using presets in Potlatch

Potlatch has an extensive set of presets (currently around 170) for many common map features that add the correct tag or tags for those features, and suggest some more you may want to add. You choose a preset from the drop-down menu just above the tag editor:

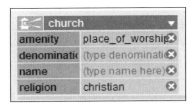

The presets are organized in groups, and you switch to or from a group by clicking on the icon to the left of the preset menu. The icon will change to show which group of presets is active, such as a car for road presets, a walking man for pedestrian-related presets, a boat for waterways, and so on. The list of presets also changes depending on whether a node or way is selected.

To apply a preset to a feature, select it, then choose the preset you want from the drop-down menu. The tags represented by that preset will be added, along with the keys of some other tags it would be useful to add. In the previous example, we've chosen the preset for a church, and Potlatch has filled in the tags for `amenity=place_of_worship` and `religion=christian` automatically, and is suggesting we add values for `name=*` and `denomination=*`. If you have the information suggested, fill in the values for those tags. You can add other tags besides the ones suggested by Potlatch if appropriate.

Using GPS traces in Potlatch

We've already seen how to load an individual trace into Potlatch, and it's also possible to see all public traces for an area. Click the GPS traces button to load them. If you hold down *Shift* while clicking the button, Potlatch will only load traces you've uploaded yourself. Traces appear as pale lines in the background.

If you load a specific trace into Potlatch, it also imports any named waypoints in the file. If you look at our example, you should see some red dots on the map. These are the waypoints from our GPS trace. Click on one, and you should see the tag editor appear, looking something like the following screenshot:

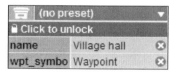

You can see the name of the waypoint, a special tag used to make it show up in Potlatch, and above these the red bar with **Click to unlock**. This means that this particular object has been locked, in this case because it's from our GPS trace, rather than from the OpenStreetMap data. Potlatch won't upload any object marked as locked to the server unless you specifically unlock it first. If we had chosen to convert the example trace to ways, these would also appear as locked objects. You can unlock waypoints and tag them correctly to create points of interest, but most of the time it's easier to use Potlatch's drag-and-drop POI icons to map these features.

Practicing with Potlatch

If you want to practice using Potlatch, but don't want to make any real mapping or risk-creating errors in the existing data, you can use the OpenStreetMap test server, which runs all the same software as the main server, but doesn't contain any data and doesn't render maps. The test server is currently at `http://api06.dev.openstreetmap.org/`, but this will change if a new version of the server software is released. Check with the API test page at `http://apis.dev.openstreetmap.org/` to see which is the latest version.

Learning more about Potlatch

Potlatch has built-in help which lists the keyboard shortcuts and offers some mapping tips. You can access this at any time using the help button at the bottom-left of the editor, but you can't keep it open while editing.

There's more extensive documentation on the OpenStreetMap wiki, at `http://wiki.openstreetmap.org/wiki/Potlatch`, including video tutorials, frequently asked questions, and the procedure for doing photo mapping in Potlatch.

Potlatch 2

There is a new version of Potlatch in development, called Potlatch 2, which has been rewritten from the ground up, and has a much richer editing environment. However, development is still at an early stage, and even when it's complete, Potlatch 2 will have greater system requirements, so it's unlikely to replace the original version of Potlatch completely and immediately. Some of the new features in Potlatch 2 include:

- Customizable tagging presets
- Customizable rendering of features in the map
- A wider range of background image types and sources

While there will be a wider range of features in Potlatch 2, the basic methods of editing won't change that much, so someone who's used to using the original version should be able to use the new version without problems.

Java OpenStreetMap Editor (JOSM)

The Java OpenStreetMap Editor, almost universally referred to by its initials, is a desktop editing application written in Java. As such, it runs on Windows, Mac OS, and Linux. The project has its own website at `http://josm.openstreetmap.de/` where you can download the latest version of the software. You'll also need to download Java if your computer doesn't already have it installed, from `http://java.sun.com/` or via a package manager on some Linux distributions.

JOSM has many features built-in, such as audio and photo mapping support, that help you turn your survey information into map data. It also supports plugins that add extra functions, such as directly gathering traces from a GPS in real time, and extra drawing tools or connections to third-party websites.

There are generally two versions of JOSM available: the **tested** version (`http://josm.openstreetmap.de/josm-tested.jar`) and the **latest** version (`http://josm.openstreetmap.de/josm-latest.jar`). The tested version is more stable, and is what you should use unless you have a particular need for some leading-edge features. The latest version is more up-to-date, but will probably contain bugs and may even crash while you're working, losing any work you've done.

Both are distributed as executable JAR files, which you can normally run just by double-clicking the file if Java is correctly installed. The JAR doesn't need installing, or to be in a particular location on your hard drive. If you have a working version of Java installed, but double-clicking the JAR file doesn't work, start JOSM with the following command line:

```
java -jar josm-latest.jar
```

Using JOSM on public computers

JOSM is packaged as a single file, which can be used without any extra installation, provided Java is working. This means you can use JOSM on a public computer quite easily, even where you don't have installation privileges. However, JOSM writes its preferences, including your username and password, to the local disk, so be wary of using it on public computers. If you do, remember to blank your credentials in JOSM's preferences before you leave the machine.

Understanding JOSM's user interface

Once you've successfully launched JOSM, you should see the main window appear. At startup, this will show the Message of the Day, downloaded from the JOSM website. The interface is a little bare until you have some data loaded, so let's do that.

Click the Open files icon (a blue folder with a file in it), or press *Ctrl+O*, to bring up the open dialog. Locate where you've unpacked the sample trace, and choose it. Once you have a GPS trace loaded, JOSM's interface gets a lot more complicated. You should see something like the following screenshot. In the center of the window is the map area, which is currently only showing our GPS trace. You should also be able to see the named waypoints in the trace, marked with crosses and labeled with their names.

Down the left-hand side of the window, a new toolbar has appeared, which contains the drawing tools and the show/hide buttons for the information panels. Across the bottom of the window is the status bar, which gives coordinates and measurements for features, as well as a hint of the current tool's function. The layout of the top toolbar is customizable, but not that of the drawing tools. JOSM's keyboard shortcuts are also customizable.

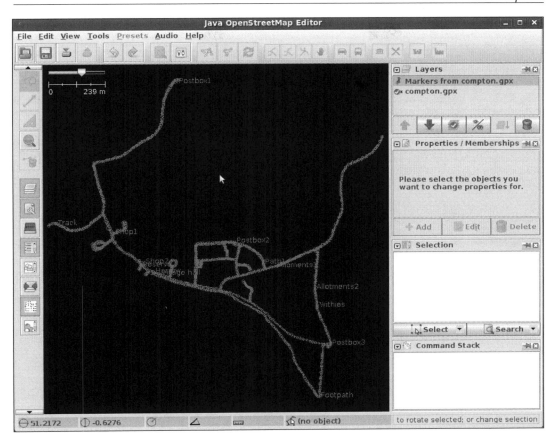

Down the right-hand side of the window are the information panels. These show information about the data in JOSM, including a list of selected objects, how the selected object(s) are tagged, who the last mapper to edit a feature was, and the different layers in your view. Each of the panels can be hidden, minimized, or floated from the main window, as follows:

- To hide a panel, click on the cross in the top right-hand corner
- To minimize a panel, click on the down arrow in the top left-hand corner
- To float a panel, click on the pushpin icon
- You can show any hidden panel by clicking its button in the left-hand toolbar

JOSM uses layers to separate different types and groups of data. Note that these aren't the thematic layers found in traditional GIS systems, where different types of features in the same set of data are separated into layers. Instead, the layers in JOSM represent either a set of OpenStreetMap data, a GPS trace or traces, or a layer added by a plugin to help mapping. You can hide or delete individual layers, reorder them, and merge layers of the same type by using the buttons below the list of layers.

You can already see two layers after loading our sample trace: the trace itself in one layer, and the named waypoints in another, called a Marker layer. Layers have actions associated with them, and right-clicking on a layer will bring up a context-sensitive menu for that layer.

Let's add another layer by downloading the existing data for the area covered by our example trace. Click on the **Download** button or choose the **Download from OSM** option from the file menu. The download dialog box will appear:

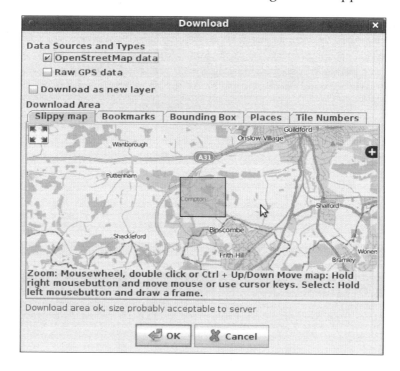

You can see the download dialog has its own slippy map, and the area covered by our trace is already highlighted. You can choose a different area to download by dragging a rectangle across the map, and you can move the map around by dragging with the right mouse button.

There are other ways to specify the area you want to download, including bookmarked areas, by entering the latitudes and longitudes of the bounding box directly (or by extracting them from an openstreetmap.org URL).

There's a limit on how large an area you can download in one request from the OpenStreetMap API of 0.25 square degrees, or roughly a 50km x 50km area in Europe. The limit prevents any one request from taking all of the server's resources. JOSM will warn you if you're trying to exceed this limit. If you need to map an area larger than this, you can perform multiple requests, and there are plugins for JOSM that will help do this.

If you need OpenStreetMap data in large quantities to use in an application, you're encouraged to use one of the many sources of bulk data, which we'll discuss in Chapter 8.

You can choose which types of data you want to download, either data, GPS traces, or both. You also have the option of downloading new data into a separate layer from any existing mapping data you may have. Let's tick the box for **Raw GPS Data**, and click **OK**. JOSM will now download the data and existing GPS traces for Compton.

Notice that so far you haven't configured JOSM with your openstreetmap.org username and password, yet it has still been able to get data from the server. Reading data doesn't require authentication, or even an account on openstreetmap.org.

During busy periods, it may take some time to contact the server and download the data. JOSM downloads GPS points in batches of 5000, again a number dictated by the server configuration. Once your download has finished, you should see a map that looks something like the following image:

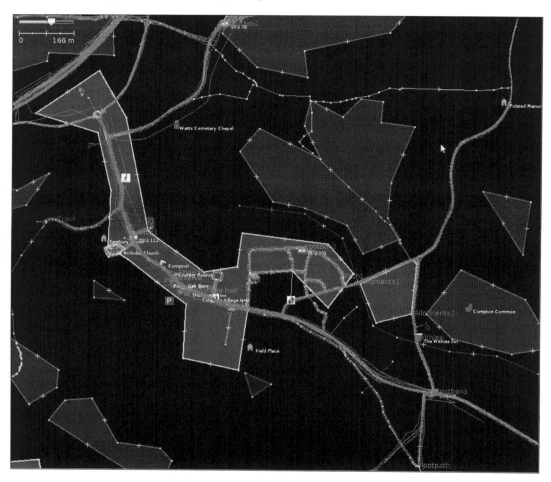

This is the completed map for Compton, and we'll reproduce most of this mapping from scratch. As the area is still being mapped, you may see more features in the data you download.

JOSM has two views:

- The **standard view** paints features in different colors and shows names of features and shades in areas. You can customize the colors JOSM uses to show features in your preferences.

- The **wireframe view** simply shows nodes and ways without any decoration. In the former view, it is easier to see which features are which while you're editing, but it is slower to redraw than the wireframe view.

You can switch between the two using the **View** menu, or using *Ctrl+W*.

You can move around the map by dragging with the right mouse button, or by pressing *Control* and the arrow keys on your keyboard. You can zoom in and out using the slider at the top-left of the map, or using your mouse wheel if you have one. There's also a zoom tool, which you select by clicking on the magnifying glass button in the left toolbar, or by pressing *Z*. Drag a rectangle over the map to zoom to that area. JOSM will make the shape of the rectangle match the current map view.

As we want to practice mapping, we need to get rid of the existing data. Go to the layers panel, which should be at the top-right of the window. Now click on **Downloaded GPX data** and click the **wastebin** button to delete the whole layer. Do the same for **Data Layer 1**. You may get prompted to save or upload your changes if you've altered any of the data you downloaded; click **Discard and Delete** if you have.

Loading images into JOSM

Now we're going to import the images in our example. In the **Layers** panel, click on **compton.gpx** to select it, then right-click to bring up the actions menu for this trace. You'll see lots of options for customizing the layer. We're interested in **Import images**, so click on that and an open files dialog should appear. Find the images you extracted earlier, select them all, and click on **Open**. You should see thumbnails of each of the images appear at various locations around the trace, and a new layer called **Geotagged Images**.

What JOSM has done is to compare the timestamp on each photo with the timestamps in the GPS trace. It has then placed the photo on the map at the closest point to the time the photo was taken. Click on any thumbnail and JOSM's built-in photo viewer pops up. You can resize the viewer to fit the shape and size of any photo, and some of our example pictures might need you to do that. Click the magnifying glass button to show the full size of the photo. You can also navigate between the photos you've loaded using the left and right arrow buttons, and have the map view center on the current photo's location.

Editing data in JOSM

To draw an object in JOSM, click on the **Draw Nodes** button in the drawing toolbar, or press *A*. Now click on the map pane, and a node should appear. Click again somewhere else, and another node will appear, joined by the first segment of a way. Carry on like this, and you can draw a whole way of as many segments as you need. Double-click to end the way, or place a single node by double-clicking.

Let's try this by going to the topmost point of the trace, which should show a named waypoint called **Postbox1** and two images. The images may appear directly on top of each other. Zoom in until the scale below the zoom slider is at around 20 m. You should see something like the following screenshot:

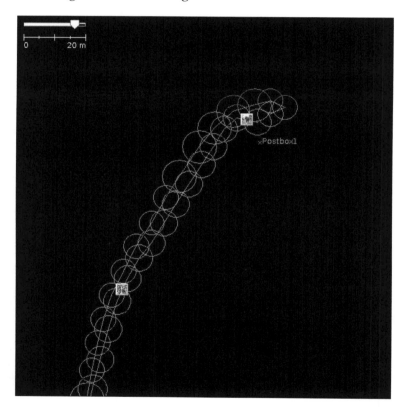

You'll see there are two lines in the trace. This is because the survey started and ended at this point, so one line represents the return journey. Having two lines means you can see the effects of the errors in your recorded position. You should also see a series of circles spread out along each line. These show the dilution of precision (DOP) of the points in the trace, giving a general indication of the possible error in position. Not all GPS receivers record the DOP, so these circles may not show up on your own traces. We're going to turn them off. Choose **Preferences** from the **Edit** menu, and the preferences dialog should appear. The **Display Settings** page should already be selected, with the **GPS Points** tab visible. On this page, you should find the option **Draw a circle for HDOP value**, which you should untick. Click **OK** to close the preferences dialog.

By default, JOSM uses separate select and draw modes, with an additional area tool. You can enter select mode by clicking on the top button in the left-hand toolbar, or by pressing *S*. In select mode, you can move around individual nodes or entire ways, or drag a rectangle around an area to select everything in that area.

The **Selection** list panel shows every object you've selected. If the feature has a name, that will be shown. If not, an icon for its type and a general description will be given. The properties panel will show the tags applied to the selected feature or features. In JOSM, you can change the tags on multiple features at the same time.

To draw features, enter draw mode by clicking on the second button in the left-hand toolbar, or by pressing *A*. Click on an empty bit of map to start drawing, and click again on the last node you drew to stop drawing. You can also stop drawing by clicking on an existing node (including one in the way you're drawing). When you stop drawing, JOSM stays in draw mode, but any further clicking on the map will create a new object. You can stop drawing by pressing *Escape*, leaving any drawing you've already done intact.

To start a new way at an existing node, select it, then enter draw mode. However, if the node you choose is at the end of an existing way, JOSM will extend this way rather than create a new one. To force JOSM to create a new way, select the node, then hold down *Alt* while drawing the first new node. To add a new node to an existing way, make sure you have nothing selected by pressing *U* on your keyboard. Now, enter draw mode, hover over the way you want to add a node to, and JOSM will highlight it. Click on the way, and a new node will appear, and then you can continue drawing normally.

JOSM has a feature aimed at speeding up mapping, called Virtual Nodes. These are enabled by default, and appear when you're in select mode as + symbols halfway along each way segment. If you click on a virtual node, it becomes a real node and you can drag it into position, and the new node is added to the way. This can make mapping a long, winding stretch of road much faster.

You can delete any feature by selecting it and pressing *Delete*, or by using the deletion tool to click on the feature you want to delete. If you delete a way, JOSM will also delete any nodes used by that way, provided they're not used by any other way or relation.

Using presets in JOSM

To add tags to a feature, you can either click the **Add** button in the properties panel, or use the **Presets** menu in the main menu bar. If you use the add button, a dialog appears, where you can type in the key and value by hand, or select them from drop-down lists based on the current data. If you type keys or values into the text boxes, JOSM will fill in the first matching value from the drop-down list. Press *Return* if this matches the text you want.

JOSM's presets are more sophisticated than Potlatch's, and for most presets will prompt you for extra information using a customized form. None of the extra information is compulsory, so if you don't have it or aren't sure what the form is asking for, it doesn't matter. Many preset forms will also include a link to the documentation on the OpenStreetMap wiki for the current tag.

JOSM also shows any presets that apply to the currently selected object at the top of the properties pane, even if you didn't use a preset to set the tags on the feature in the first place. You can click on any preset description here to show the form associated with it. In the following screenshot, you can see the presets associated with the tags `amenity=place_of_worship` and `tourism=attraction`, as applied to the **Watts Cemetery Chapel**. The preset form for a Place of Worship is also shown.

Adding your account information to JOSM

Before you can upload data to OpenStreetMap using JOSM, you need to enter your account details. There are two ways for JOSM to authenticate itself to the OpenStreetMap server:

- Username and password authentication. This uses the same username and password that you use to log into openstreetmap.org and is simple to set up, but transmits your details across the Internet in plain text, which could be a security risk in some situations.

- OAuth is more secure, but needs you to grant JOSM access to your account at openstreetmap.org before you can edit.

For either method of authentication, you need to open JOSM's preferences by choosing **Preferences** from the **Edit** menu, or pressing *F12*. When the preferences dialog appears, choose the **Connection Settings** panel—the second tab in the list:

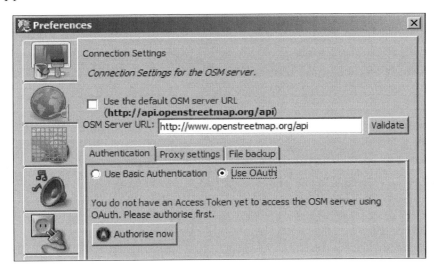

Let's use OAuth to authenticate with the server, and we're also going to use a method that doesn't involve sending our password over the Internet in plain text. We need to grant access to JOSM to do this by carrying out the following steps:

1. Make sure you're logged into openstreetmap.org in your browser.

2. In the **Authentication** tab, click on **Use OAuth**.

3. Click **Authorise now**. The OAuth dialog box should now appear.

4. In the top drop-down list labeled **Please select an authorization procedure**, choose **Semi-automatic**.

5. The layout of the box will change; click **Retrieve Request Token**. JOSM will now open openstreetmap.org in your browser, showing a page titled **Authorize access to your account**, as seen in the following image:

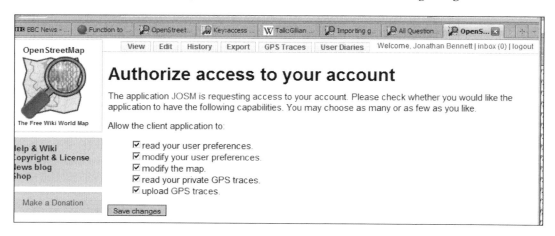

6. Leave all the permissions listed on this page ticked, and click **Save changes**.
7. Now switch back to JOSM and click **Retrieve Access Token**.
8. Once JOSM has the token, click **Test Access Token** to make sure it's working.
9. Assuming all is well, click **Accept Access Token**.

JOSM can now upload data to OpenStreetMap using your account. You can revoke this access at any time from your account page on openstreetmap.org, but it's unlikely you'll ever need to do so.

Extending JOSM with plugins

You can add extra functions to JOSM by using one of the many plugins available. These add features, such as making it easier to draw complex shapes, and tagging checkers and connections to other OpenStreetMap tools. You can add plugins through the **Plugins** panel in the **Preferences** dialog. Each plugin has a short description and a link to further information on the OpenStreetMap wiki.

Learning more about JOSM

JOSM has a built-in help viewer, but this just pulls pages from the JOSM wiki at `http://josm.openstreetmap.de/wiki`, so you can't use this if you're offline. You can view this documentation in your normal web browser if you like.

Apart from the documentation at JOSM's own wiki, there's information in the main OpenStreetMap wiki at `http://wiki.openstreetmap.org/wiki/JOSM`. You're more likely to find contributions from JOSM users in the latter location.

Merkaartor

Merkaartor is another desktop editing application, this one written in C++ and available for Windows, several versions of Linux, and a pre-release version for Mac OS X. It's a native application, which means you need to install it before you can use it, but it can be used for offline mapping.

Merkaartor is much closer in appearance and use to a traditional GIS system than the other OpenStreetMap editors. It has a high-quality rendering, multiple layers, sophisticated management of map changes, and drawing tools. However, it's not extensible in the same way as JOSM, and some features such as audio mapping aren't included. Along with editing data, you can use Merkaartor to render maps, including customizing the style.

You can download the latest version of Merkaartor from `http://merkaartor. be/`, where you'll also find installation instructions for your platform. The other documentation for Merkaartor is kept in the OpenStreetMap wiki at `http://wiki. openstreetmap.org/wiki/Merkaartor`, and there are links to the relevant pages on the Merkaartor website.

After following the instructions for downloading and installing Merkaartor for your platform, you can launch the application, through the Start menu on Windows, or the Applications menu in Ubuntu Linux. For other Linux distributions, you may have to use the command line. The user interface is straightforward, and has some similarities with JOSM. There's a toolbar across the top of the Window, information panels, called Docks in Merkaartor, down the right-hand side, and a status bar at the bottom.

Let's get some data into the user interface, so choose **Open** from the **File** menu, or press *Ctrl+O*, to bring up the open files dialog. Select our example trace and click Open. You'll see the trace appear in the map area against a shaded background. As with JOSM, this indicates the extent of the downloaded data which, as we haven't downloaded any yet, means the whole map. Let's do that now. Click the **Download** button on the toolbar, and the download dialog will appear.

Merkaartor's download dialog contains many of JOSM's features, but in a single view. There are bookmarked locations, the ability to enter an openstreetmap.org URL, and a slippy map view to choose the area to download.

To bookmark your current view in Merkaartor, select the
View menu, then the **Bookmarks** sub-menu, and click on
Add. Type in a new name for the bookmark, or overwrite an
existing one by choosing its name.

By default, the current map view is selected, so click **OK** to download it. After you
have done so, you should see something like the following screenshot:

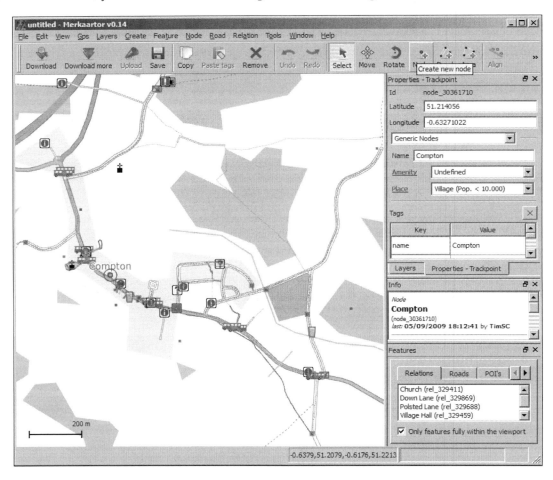

The map area shows the current data in a style similar to the main Mapnik rendering on openstreetmap.org, although there may be slight differences. Merkaartor can show data using a number of styles, and you can edit the styles or create your own. Points of interest are shown using specific icons where possible, or using an information icon for other point features. You move the map around by dragging it with the right mouse button or using the arrow keys, and zoom using the +/- keys or the mouse wheel.

Across the top of the window is the toolbar where some standard buttons and Merkaartor's drawing tools are found. Unlike the other two editors, Merkaartor uses different tools to draw point features, roads and other linear features, and area features. The Node tool only draws single, unconnected nodes. The Road tool, despite its name, is used to draw any linear feature mapped with a way. The area feature draws a closed way to represent an area feature. You stop drawing in any tool by pressing *Escape*.

Down the sides of the map area, you'll see the Docks, as they're known in Merkaartor. These panels provide information and allow you to edit the map. Each of the docks can be moved to any location at either side of the window, stacked on top of each other, or floated from the main window.

Merkaartor manages the data you download, create, and upload very carefully using separate layers. The data you download is always kept in its own layer and never modified. Instead, your edits are kept in the Dirty layer until you upload them, then they're moved into the uploaded layer. You can ignore all this while you're editing, but if you need to see just your changes, this system allows you to do it.

We want to do our mapping from scratch, so we need to discard the downloaded data. In the **Layers** panel, you should see a layer for our download, with a name matching the time and date it was downloaded. Right-click on this and choose **Close**.

Loading images in Merkaartor

Merkaartor includes support for photo mapping. To use this, we first have to show the Geo Images dock, by choosing it from the **Docks** submenu of the **Window** menu. Now bring up the **Open** dialog.

You need to load images in Merkaartor before making any changes to the map. If you try to bring up the Open files dialog while you have unsaved or uploaded changes, you'll get a message asking if you want to discard your changes.

Now, select all of our example images, and click **Open**. You'll be prompted to select the layer with which you want to associate the images. You should choose the layer titled **compton.gpx – Track 1**, and you'll get another prompt warning you that the layer is read-only, and asking if you want to make it writeable. Click **Yes**. Yet another dialog will appear, asking you to specify the time offset between the GPS trace and the photos. For our photos this is zero, so just click on **OK**. You may also get another prompt warning you that a large time difference between the timestamp on a photo and the GPS trace exists. This warns you that the positioning of a photo may not be accurate, but in our case it's not a problem, so just click **Yes to All**. Merkaartor should now import all the photos.

Merkaartor shows the location of each photo by drawing a rectangle around the trackpoint matching its timestamp. Click on this rectangle to show the photo in the Geo Images dock. If you have space on your desktop, floating the Geo Images dock and enlarging it will allow you to see more detail in each photo.

Editing data in Merkaartor

As we've seen, Merkaartor differs from the other two editors we've looked at, in that it uses different tools to draw different types of features. It also has separate select and move tools, which can prevent you from accidentally moving features. However, you can disable this behavior if you like and have a single tool for both functions. Choose **Preferences** from the **Tools** menu, select the **Visual** tab when the **Preferences** dialog appears, and untick **Separate Move mode**.

To select a feature in Merkaartor, hover over it and a blue highlight should appear. For nodes, this will be two circles around it, for a way, the center line of the way, and for an area, its outline. Click on the feature, and it will get a thick outline. If you select a node, all the ways it's a part of will also be highlighted using a dotted line.

Merkaartor shows the tags applied to a feature in its **Properties** dock, as well as the object ID, latitude and longitude for nodes, and any presets that apply, as shown in the following screenshot:

In the preceding screenshot, we have the properties for The Harrow Inn—one of Compton's pubs. Merkaartor allows you to edit the coordinates of any point directly, although this isn't normally necessary.

Merkaartor's preset system is built into the properties dock. It's less sophisticated than the presets in the other two editors, and has far fewer options available. The major types of features such as roads, settlements, and amenities are present, but little else. Fortunately, the text boxes for entering tags manually have a drop-down list based on the current dataset, so you can still use common tags without always having to type them out in full.

Relations are shown in Merkaartor using a dotted rectangle around all of the relation's members. These can extend outside the area you've downloaded, but you can also get Merkaartor to download all relation members by ticking the **Resolve all relations** box in the download dialog. To select a relation, click on its outline, and its members and tags will appear in the properties dock.

Uploading edits to OpenStreetMap

To upload your edits, you need to add your account details to Merkaartor's preferences, by carrying out the following steps:

1. Open the **Preferences** dialog in the **Tools** menu.
2. Switch to the **Data** tab.
3. Enter your username and password into the fields in the form, leaving the **Website** field set to **www.openstreetmap.org**.
4. Click **OK**.

You're now able to upload the edits you've made.

Learning more about Merkaartor

Merkaartor's documentation is held in the OpenStreetMap wiki, and there's a dedicated Merkaartor mailing list at merkaartor@openstreetmap.org.

Summary

You should now be familiar with the basic operation of OpenStreetMap's three most popular editors:

- Potlatch
- JOSM
- Merkaartor

and hopefully have an idea which will suit you best.

You should know how to:

- Download and run each one
- Load in a GPS trace and load images where appropriate
- Operate the basic drawing tools in each one

Next, we'll look at some specific mapping techniques that apply to all editors.

6

Mapping and Editing Techniques

Although there are differences between the various OpenStreetMap editing applications, the techniques used to represent physical features using the data structures we covered in Chapter 4 apply to all of them. In this chapter, we're going to look at how drawing nodes, ways, and relations, then tagging them, turns them from simple data structures into representations of geographic features. We're going to cover:

- Drawing ways using GPS traces for guidance
- Adding points of interest
- Modifying existing features
- Drawing complex features
- Using editor presets to make tagging common features easier
- Manually tagging less common features
- Finding undocumented tags already in use in OpenStreetMap

By the time we've finished, you should be comfortable using any editor for OpenStreetMap to add data to the map, tagging it in accordance with current community practices, and uploading it to the server. We'll work through our example trace and the accompanying photos in the same way a mapper would after a survey.

Drawing and tagging features

We're going to draw the roads and paths in our example first, then add the point features after that. Between the GPS trace, the named waypoints, and the photos, we have enough information to start mapping. Normally, you should download any existing data from OpenStreetMap and add your mapping to that, but we're going to assume that this area has no previous mapping at all.

We're going to start by drawing the roads on the map. You can use any of the editors to practice, but please remember not to upload your data to the server, or you will duplicate existing data. Choose whichever editor you like the look of, or even repeat this exercise with each editor to get a feel for them.

Whichever editor you're using, make sure you have our example trace loaded, and for JOSM and Merkaartor, load the images. You can still use the images as a reference if you're using Potlatch, but if you were doing this in a real survey, you'd have to take a note of which photos were taken where.

We'll start where the survey itself starts and ends, at the northernmost point of the trace, by a named waypoint called **Postbox1**. The trace leads southwards from here, and reflects the shape of the road. You can see two lines in the trace; this is because we returned to the starting point after surveying the rest of the village. This isn't a requirement of any survey, but it helps to get more than one trace of each road over time to increase the accuracy of the map.

Click between the two lines at their northern end, and then keep adding new points, following the lines. How often you place a node is a matter of judgment. You need to reflect the general shape of the GPS trace without using too many nodes. Nodes should be at least 5 meters apart unless you're mapping a very tight corner, or there are other features on the road that need mapping. I've used 18 nodes for this section. Once you get to the end of this road, where the trace turns sharply south, place a node at the corner, then stop drawing by clicking the node again in Potlatch and JOSM, and by pressing *Escape* in Merkaartor. In Potlatch, the result should look something like the following:

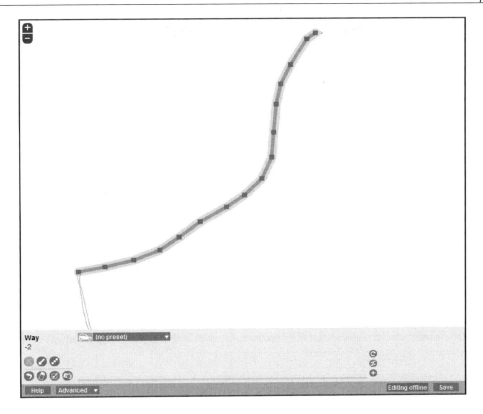

Now we have an untagged way, which shows us physical location and nothing else. At this stage, we could turn this way into a road, a footpath, a river, or any other linear feature. We're actually going to make it an unclassified road. What constitutes an unclassified road varies from country to country, but in the UK this is generally any through road without a higher classification, which is the case for this road. The road's name is **Down Lane**, which can be seen in the image `121120009090.jpg`. All three editors have unclassified roads in their presets, so let's use that:

- In Potlatch, make sure you have the car icon shown in the presets, then choose **unclassified road** from the drop-down list. Enter the name in the blank **name** tag Potlatch has added.

- In JOSM, go to the **Presets** menu and choose **Highways | Streets | Unclassified**. Enter the street name in the dialog box for the preset.

- In Merkaartor, choose **Unclassified road** from the **Highway type** drop-down in the **Properties** pane, and enter the name in the **Name** field.

You should see the rendering of the road change as you add the tag. Even in JOSM's wireframe mode, untagged ways are drawn in a different color than tagged ones.

That's it; you've created a road. If you uploaded this data to OpenStreetMap (although you shouldn't because you'd be duplicating existing data), your new way would be assigned an ID, added to the database, and rendered on the two main rendering on openstreetmap.org after a short delay. It would also appear in any of the downloads of the area's data from the OpenStreetMap API.

How often should I upload edits?

How often you should upload your work is a matter of judgment, but there are some things to bear in mind. You should upload often enough that you don't risk losing your work, but not so often that it's impossible to trace groups of related edits. If you're working on an unreliable machine or connection, upload more often. If you're editing the same area as other users, your edits could conflict with theirs, and uploading often will help prevent this.

The next feature to draw is the main road that runs through the village. In JOSM and Merkaartor, you can look at the photo that's just to the south of the turn. This shows the name of the road as The Street.

We need to draw the road starting at the last node we drew of the last road, but we also need to make sure that we have a separate feature for it. To draw a new line from an existing node, do the following:

- In Potlatch, select the way we drew previously, then shift-click on the last node to create a new way. Now draw as normal.

- In JOSM, select the last node, then hold down *Alt* and place the first node of the new way. Now release *Alt* and carry on drawing.

- In Merkaartor, choose the **Road** tool, then hover over the end node of the previous way. A circle should appear around the node. Click on the node and draw as normal.

The trace has two lines for this road along two thirds of its length, and then they diverge. Follow the northernmost of the lines, which ends near a waypoint named **Postbox3**. You can save yourself some time by placing nodes where there are side-roads, so you can start drawing from them later.

Once you've drawn the way, we need to tag it. Let's use the presets again, but this time choose **Secondary road**. Add the name **The Street**. In Potlatch and JOSM, we now have an extra tag suggested in addition to the name—ref=*. Secondary roads are expected to have a reference associated with them, and in the case of this road, it is B3000. You'll find this reference on signs along the route, and again precise use of the ref=* tag on roads varies from country to country. In Merkaartor, you'll have to enter the ref=* tag by hand.

Now let's draw the side roads. The first of these heads westwards from The Street, around 200 meters south of the junction with Down Lane. The image by the junction shows its name as Eastbury Lane. It also has houses along its length, so we'll tag it as a residential road. This is in the presets for all our editors, so draw the road, tag it as residential, and name it.

However, there's a detail we've missed. Towards the end of the trace, you can see a named waypoint called **track**. At this point, the road turns into an unsurfaced track and continues for a short distance before coming to an end. We need to tag this section of the road differently, but fortunately that's not a problem. We just need to split the way into two at that point, and change the tags on the end section. To split the way, make sure you have a node in your way where the waypoint is, and use the editor's split action, as follows:

- In Potlatch, click on the way to select it. If you don't already have a node at the point we want to split at, shift-click to add one. Click on the node at the split point, which should then be highlighted, and click on the scissors button to split the way.

- In JOSM, select the node at the split point and choose **Split Way** from the **Tools** menu.

- In Merkaartor, select the node at the split point, then choose **Split** from the **Road** menu.

Now, re-tag the new way at the end as a track. Its rendering should change as a result. In Merkaartor, the result should look something like the following:

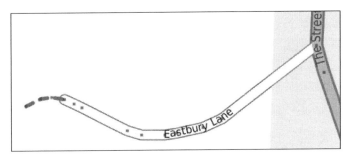

Let's move on to our next side road, this time on the east of the main road, starting near a named waypoint called **Shop1**. This is just a lane leading down to the car park of the village church (we'll map the church after we've finished the roads). Lanes like this are usually tagged as `highway=service`, denoting a road used just to access a particular property. Despite the traces for this side road wavering slightly, the lane is perfectly straight, so draw it using a single section of way. Occasionally, you'll encounter parts of a GPS trace, where your own memory or some record you've made is a better guide of how to map a feature. Tag the way as a service road, either using presets or manually.

Drawing area features

Now let's draw the car park at the end of the lane. You can see a rough shape drawn by the trace, but the parking itself is rectangular. Even if the shape of the parking area was slightly irregular, it's best to map it as a regular polygon at first, particularly if it's a small area like this one. Again, this can be refined later on by yourself or another mapper. To draw an area for the parking:

- In Potlatch, select the node at the end of our service road, shift-click on it to start a new way, and draw the rectangle by hand, ending on the same node you started with. Potlatch stops drawing when you close an area in this way. You'll have to adjust the shape of the area by hand.

- In JOSM, select the node at the end of the service road. Then select the way tool, and draw a single line at a right-angle to it, holding down *Alt* to start a new way. Now select the area tool, which has a set square icon. Click and drag the line you've just drawn with this tool, and JOSM will draw a rectangle for you.

- In Merkaartor, choose **Rectangle** from the **Create** menu, and draw a shape of roughly the right size, in roughly the right place. You're forced to draw the rectangle aligned with the grid, but once you've finished drawing, you can select the rectangle, then use Merkaartor's rotate tool to rotate and size it until it fits our car park. Move the rectangle so its bottom node overlaps the end node of the service road. Select them both and choose **Merge** from the **Node** menu.

Let's tag the area manually, using the tag `amenity=parking`. This just involves typing the key and value into the text fields in the editor.

Drawing ways with loops

Our next road to map is around 300 meters further along the main road on the northern side. You should be able to see a loop at the end, and a photo of its name sign at its junction with The Street. The road is a residential street that loops around onto itself to form a small square. Although we could draw this arrangement as a square and a separate section of road, it's better if you use a single way to represent a road where possible. The simplest way of doing this is to start at the junction of the main road, draw the first section of road, then half of the bottom section of the square, the other three sides then finish back on the node halfway along the bottom side. You should end up with something like the shape shown in the following image.

This shows the street using JOSM's wireframe mode, with its segment order number feature enabled. You don't normally need this feature enabled (and no other editor has an equivalent feature), but it's sometimes useful in deciphering the structure of complex ways. You can enable it in the **OSM Data** tab of the **Display** preferences.

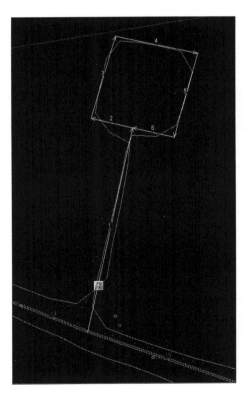

Tag the way as `highway=residential` and `name=Fowlers Croft`, taking the name from the photo.

Mapping residential streets and complex streets

Next, we're going to map a small but relatively complex set of residential streets. These start at the next junction eastwards from Fowlers Croft. The image at this junction (`12112009101.jpg`) tells us that this particular street is named **Spiceall**, and also that there's a telephone box near the junction. We'll map the telephone box later. Spiceall extends all the way along the trace from the junction to the waypoint named **Path1**. Draw a way to represent this. There's also a side street two-thirds of the way along the first section of the street. However, there are no name signs at the entrance of this road. A quick inspection of the house numbers tells us that this side street is part of Spiceall, so we now have a new situation to deal with. We could map the two parts of the street separately, adding a `name=*` tag to each. This would produce a reasonable looking map, but would also result in there being two features in the OpenStreetMap database called Spiceall, whereas there's only one street with that name; it just happens to be non-linear. To show how the two parts of the street are connected, we're going to use a relation.

Draw a way for the side street, which ends where another part of the trace almost joins from the north. You should be able to do this with just a single segment. Now we need to create a relation and add the two ways to it, as follows:

- In Potlatch, select one of the ways, and click on the relation button (a chain icon) at the bottom right of the editor. At the prompt, choose **Create a new relation** and click **OK**. The relation editor should appear.

- In JOSM, make sure you have the relations panel showing, clicking its button in the left-hand toolbar if necessary. Now select both ways, and click the **Create relation** button at the bottom of the relations panel. JOSM's relation editor should appear.

- In Merkaartor, select both ways, then open the **Create** menu and choose **Relation**. The **Properties** panel will change to show the new relation.

Now that the relation is created, we should tag it. The first tag we need to add is a `type=*` tag to identify what this relation represents. In this case, it's a collected street, which is identified using a `type=collection` tag. Add the `highway=residential` and `name=Spiceall` tags to the relation, which you'll have to do manually, as there's no preset support for relations.

Renderer support for relations

There's a varying level of support for relations in the two main renderers used on openstreetmap.org, and not all relation types will show up. At the time of writing, the preceding example would fail to render, as the renderers hadn't been patched to support the `collection` type. In these circumstances, the individual ways are tagged with `highway=residential` to make them render, but this is a short-term hack.

Leading off Spiceall is another complex street. From the image at its northernmost end (`121120009103.jpg`), we can see its name is Almsgate. It extends from Spiceall directly south to the junction with another road, yet to be mapped, and has its own side-street. Map and tag Almsgate in the same way as Spiceall. By the time you've finished, you should have a set of streets shaped like the following, as seen in JOSM:

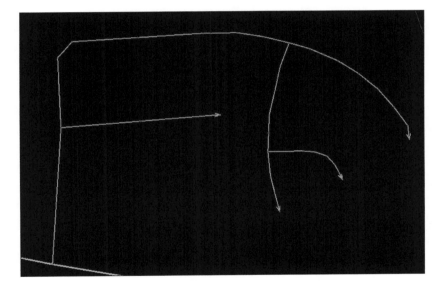

There are still some sections of trace in this area we haven't mapped yet that represent footpaths. These connect the end of Spiceall's spur with Almsgate and the other branch of the street. The former of these is just a straight line between the end of the spur and Almsgate, which we'll map using existing nodes, as follows:

- In Potlatch, select the end node of Spiceall by clicking on the way, then shift-click on the end node, and Potlatch should start drawing a new way. Now move across to Almsgate and hover over that way. Potlatch will show its nodes. Click on the one close to the end of the trace, then click again to stop drawing.

- In JOSM, select the end node of Spiceall, then press *A* to start drawing, then hold down *Alt* while clicking on a node in Almsgate. Click on the node again to stop drawing.
- In Merkaartor, choose the **Road** tool, then click on the end node of Spiceall. Now click on a node in Almsgate, and press *Escape* to stop drawing.

You now have a new way which you should tag as `highway=footway`, which is in the presets in all three editors. There's another path to map leading north from the end of the Spiceall spur. The trace for this path looks quite messy, and in real life it makes a few 90-degree turns between the two roads. Draw the best fit you can, remembering that you can always do another survey at a later time to increase the accuracy of the map. Adding the path using a representative shape still lets people know there's a path there they can use. Use the same technique as with the previous path.

You could draw these footpaths without reusing the existing nodes, and the rendered map would look very similar. However, routing applications wouldn't know that the footpaths were connected to the roads, and wouldn't be able to calculate a route using our new footpaths. It's important that you connect any paths and roads using shared nodes where you can travel between them, otherwise, routing algorithms won't work.

Our next road leads off The Street and heads northeastwards past our residential road and comes to an end some distance from the village. The photo (`121120009108.jpg`) tells us it's named Polsted Lane. There are a few houses on this road, but much of its length is bordered by woods or fields. We could classify it either as a residential road or an unclassified road, but we'll choose `highway=unclassified`.

Barriers and gates

Once you've drawn the road, join the ends of Spiceall and Almsgate to the road using footpaths. Although we don't have a trace between the end of the Almsgate spur and Polsted Lane, we have a photo (`12112009104.jpg`) showing the short section of path. It also shows a barrier on the path. All barriers in OpenStreetMap use the tag `barrier=*`, but not all barrier types are included in the editor presets. Even if there is a preset, it can be unclear which preset refers to which kind of barrier.

For any feature you can't find a preset for or are unsure which to use, you can look the feature up in the OpenStreetMap wiki. The fastest way of finding information on a tag is to search for the key. In most cases, this will lead directly to the right page, and in our case it does. Enter **barrier** into the search box in the left-hand sidebar, and you'll be sent to the page for `barrier=*`. A table on this page lists most of the types of barriers that have been mapped and documented by other mappers.

We can see from the photos on the page that this kind of barrier is tagged as `barrier=cycle_barrier`, which is in the presets for Potlatch and JOSM, or you can just tag it manually. Place a node halfway along the path and tag it to represent the barrier.

We need to complete the roads in the village. There's one between The Street and Polsted Lane, which the photo at its southern end (`12112009110.jpg`) shows, called Withies Lane. Again, the ends need to share nodes with the two roads we've already mapped, so reuse nodes if they're in the right place, or add new ones to the existing ways.

Next, there's another road leading south from The Street, just to the west of the end of Withies Lane. The name is given in photo `121120009115.jpg` as The Avenue. This road actually extends past where our survey ends, which will be a regular occurrence in real surveys, as you'll have limited time and will have to limit the area you survey. It's perfectly OK to leave a road unfinished in this way, as either you or another mapper can complete it at a later stage. If you want to let other mappers in the area know that this road needs finishing, you can add the information in a `note=*` tag.

The last section we have to map is a footpath leading from The Avenue back to The Street. This starts at the waypoint named Footpath, and cuts across to meet The Street between Spiceall and Fowlers Croft. Although the path is fairly straight, there are a few bumps in it, and it's a matter of judgment how closely you follow these bumps in your mapping. Even with a good trace like this, with a low dilution of precision, you can't be sure it's a perfect representation of the path's course. Using a rough approximation to the path won't get people lost, provided you're never more than a few meters out of the true position. Draw the path and tag it as `highway=footway`.

Points of interest

Now that we've added all the roads, we'll add the rest of the features we surveyed. Again, we'll use the editors' built-in presets, but we'll also need to use some other resources to get the right tags for some of the features.

We'll map the features in the order they were encountered during the survey, so we'll go back to the start of our trace. Our survey starts outside the Watts Gallery Estate, which we can see from our photo of the sign at its entrance (`12112009087.jpg`) has an art gallery and a cafe. For now, it's enough to record these features using a single node each, roughly in the right place. This will be accurate enough for people to find the gallery using OpenStreetMap. The extent of the estate and its precise positioning can be drawn later from a separate survey, thanks to the wiki-like nature of the OpenStreetMap API.

The estate is to the north-east of our starting point, so place a single node around 30m away for the gallery, and another next to it for the cafe. We could add the tags for both features to a single node, but this would make rendering images for both features difficult. Tag one node as the gallery by selecting it, then using a preset:

- In Potlatch, click the preset group icon until you get the camera, for tourism presets. Choose **attraction** from the drop-down menu.

- In JOSM, choose **Travel | Tourism | Attraction** from the **Presets** menu.

- In Merkaartor, tag by hand as `tourism=attraction`, as it doesn't have tourist attractions as a preset.

Tag the other node as `amenity=cafe`, which appears in the presets for all editors, or by hand if you like.

Just outside the gallery is a postbox. Its location is marked with a named waypoint, so place a node in that location and a tag as `amenity=postbox`. The preset in JOSM suggests you add an extra `operator=*` tag, while the preset in Potlatch suggests you add a `ref=*` tag. Merkaartor doesn't suggest any extra tags, so what's correct? They both are, as it turns out. The wiki page for the `amenity=postbox` tag suggests that both these tags as useful information, and further suggests recording the collection times in OpenStreetMap.

If you look at the photo of the information on the postbox itself (`121120009086.jpg`), it tells you the collection times and a reference number—GU3 78. While the reference number may not mean much to a person using the postbox itself, it's a unique identifier for that box, and may prove useful for some applications at some point, so we'll add it to the database. The `ref=*` tag is used for this, so tag the postbox as `ref=GU3 78`.

Record as much information as possible

Information like this, which may not seem immediately useful, can be added to OpenStreetMap without causing any harm. If it has a practical application, then it's already present and you won't have to re-survey an area just to gather the extra tags. OpenStreetMap data has been used to create specialist maps that make use of information in unpredicted ways, so any verifiable details of a feature should be recorded.

If it turns out that it's not useful information, the only effect is that it will take up some extra disk space.

Our next point of interest is just down the road, located using a photo, `12112009089.jpg`. There's no waypoint to show its location, but thanks to being able to place photos using their timestamp, we still know where the feature is. The Watts Cemetery Chapel is just to the east of the road. We haven't surveyed the whole cemetery grounds, so just add a node to the east of the road and use the presets to tag it as a place of worship, setting **religion** to **christian**. Also tag it as `tourism=attraction`, as people visit the chapel because of its association with the artist, Mary Watts.

Moving onto The Street, there's a waypoint named **Shop1**, but no other information accompanying it. What happened here is that a photo meant to show us what the shop was named and what it sold hasn't worked, for whatever reason. Problems like this can occur during surveys, and unless you double-check everything, which often isn't practical, you'll miss a few details. If you have no other information and can't remember what the feature was, all you can do in these circumstances is try to find a generic tag for the feature you're trying to map, or resurvey at some later date.

There's another post box just by this waypoint, shown in the photo `121120009094.jpg`, which you can map in the same way as the first one that we encountered. There are another two postboxes in the village, both marked with both waypoints and photos. You can skip ahead and map these now, or work through the features in order. It's up to you in which order you map features.

Across the road, there are a couple of photos to mark more features. The first, in photo `121120009095.jpg`, is a memorial. The tag for a memorial is `historic=memorial`, but this is in the presets in JOSM, but not Potlatch or Merkaartor. Place a node and tag it with `historic=memorial`.

Mapping overlapping features

Alongside the photo of the memorial is another one (`12112009096.jpg`) of the sign at the entrance to our next feature—the village church. The name, seen in the photo, is St Nicholas Church. The trace leads south-west from the photo and then completes a couple of loops. The inner loop represents the footpaths around the church, and the outer loop the perimeter of the church yard. The church itself is at the center of the site. You should map the footpaths using two ways, which connect at each end, tagged as `highway=footway`. One of the ways should extend all the way to the main road, past the memorial, so pedestrian routing applications can get people to the church.

We can also add the church itself as a node at the center of the loops. All three editors support places of worship in their presets.

- In Potlatch, select the bank of presets represented by a lighthouse icon, then choose **church** from the drop-down list.

- In JOSM, choose **Presets | Man Made | Buildings | Place of Worship**. Choose **Christian** from the **Religion** drop-down.

- In Merkaartor, choose **Place of Worship** from the preset drop-down, then choose **Christian** from the **Religion** drop-down.

For JOSM and Merkaartor you can set the denomination (in this case Anglican) using the presets, but in Potlatch you'll need to fill in the field by hand. Add the name exactly as seen in the photo.

Next, we need to map the church yard. This is done using an area, in a similar manner to the nearby car park we mapped earlier. However, the situation is slightly more complicated, because we have to map the church yard around the existing church and footpaths.

Draw a closed way around the outer loop of the trace, but be careful not to let it join any of our existing ways. In particular, avoid joining the way to the footpath you've already drawn.

Normally, areas are mapped independently of any features within them. The exception to this is where a highway of some type passes through a barrier around the perimeter of an area. While there is a wall around the churchyard, we're not going to map this, as we're not covering buildings and other structures in our simple survey.

Once the way is complete, tag it as `amenity=grave_yard`. There's a preset for this in JOSM under **Geography | Land Use**, despite not using the `landuse=*` key. The other editors don't have a preset for this. There is also another separate tag `landuse=cemetery`, which is used for burial places that aren't associated with a particular place of worship. Again, you could read the documentation in the wiki for these tags to find out which is most appropriate for the area you're mapping.

Once you've finished drawing the church and yard, you should have something like the following in JOSM's wireframe view:

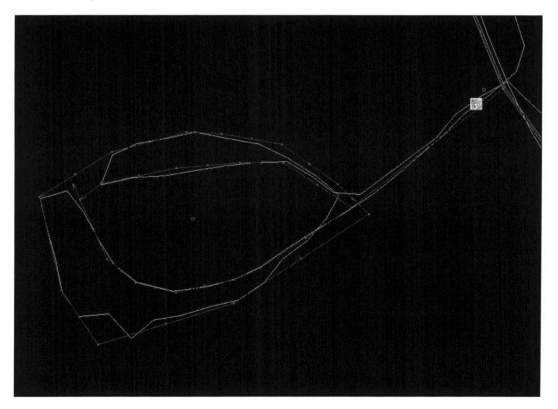

You can see the mapping hasn't followed the trace precisely. This is because the trace doesn't accurately reflect the shape of the churchyard and paths, partly through errors in the recorded positions, and partly because it wasn't possible to completely follow the perimeter of the area. It's fine to generalize mapping like this as long as you don't introduce glaring errors.

Other civic amenities

Further south along the road is a named waypoint Shop2, this time accompanied by a photo, `121120009098.jpg`. The photo shows the shop's name as Country Rustics, and its telephone number is visible. The shop itself appears to sell antiques. Add a node and tag it with the name, a `shop=antiques` tag and a `telephone=*` tag for the number. Although the telephone number may change at some point, OpenStreetMap can be updated if it does.

Our next point of interest is across the road, shown in photo `121120009097.jpg` as the McAlmont Reserve — a small nature reserve and public hall. We're going to ignore the orchard part of the reserve in our survey, but mark the location of the public building — Pucks Oak Barn — on the map. A search for "public building" in the OpenStreetMap wiki leads to the tag `amenity=public_building`. Add a node at the end of the short spur of trace by the photo, opposite the shop, and tag it `amenity=public_building`, and a `name=*` tag.

A waypoint named **Harrow** with an accompanying photo (`121120009099.jpg`) marks the location of one of the village's pubs — The Harrow. Map this as a node to the south of the main road, and use the presets to tag it, as follows:

- In Potlatch, click the preset icon until you get the camera, then choose **pub** from the drop-down list.
- In JOSM, choose **Travel | Food + Drinks | Pub** from the **Presets** menu.
- In Merkaartor, choose **Pub** from the **Amenity** drop-down list in the **Properties** panel.

Add the name, and it's just possible to see the pub's telephone number in the picture, so tag that as well. There's another pub — the Withies Inn — located by the waypoint named Withies.

Just along the road from The Harrow is a waypoint named Village hall. Searching the wiki for "village hall" produces the tag `amenity=townhall`, which despite the name is intended for public halls in any size of settlement.

In the photo showing us the name for Spiceall, we could also see a phone box, so add this as a node at the corner of Spiceall and The Street, on the western side of Spiceall. Tag it as a phone box, as follows:

- In Potlatch, select the Postbox icon in the presets, then choose **phone box** from the drop-down list.

- In JOSM, choose **Man Made** | **Amenities** | **Telephone** from the **Presets** menu.

- In Merkaartor, choose **Public Telephone** from the **Amenity** drop-down.

Finding undocumented tags

Next, we come to Compton Village Club on Spiceall. This is a private members' club where general public entry isn't permitted, so it's not a pub like the previous points of interest we've mapped. None of our editors have a suitable preset, so we have to look elsewhere to find out how to tag this feature.

The next place to look for information is the OpenStreetMap wiki. As facilities such as pubs and clubs are normally tagged using the `amenity=*` key, we look at the wiki page for that, namely, `http://wiki.openstreetmap.org/wiki/Key:amenity`. At the time this book was written, there are entries on this page for pubs and nightclubs, but this isn't a nightclub in the normal use of the word. A more general search of the wiki for "social club" or "members club" doesn't produce any helpful results, so it looks like no one has documented a tag for a social club.

This doesn't mean no one has used such a tag, just that no wiki page has been written, so we need to use other tools to find out what other mappers have done.

In this case, we're going to use a third-party web-based tool called OSMdoc (`http://osmdoc.com/`) that analyzes the tags in the OpenStreetMap database and produces statistics showing what tags are used, how often, and on which primitives.

Open the home page in your browser and you'll see a table of information, as follows:

OSMdoc

Tags						

Show 25 entries   Search: [ ]

				Usage		
Key	Values	Total	Nodes	Relations	Ways	Changesets
source	11675	200803210	195867526	40317	4895246	121
tiger:tlid	41216286	183689694	169947429	0	13742265	0
tiger:county	4362	179525846	166135561	0	13390285	0
tiger:upload_uuid	4558	179144251	165777024	0	13367227	0
created_by	4484	50131444	40676020	49731	8215685	1190008
highway	1437	25615942	660797	338	24954807	0
name	5486118	16720905	3090860	95468	13534577	0
tiger:source	120	13774780	32241	0	13742539	0
tiger:cfcc	586	13747973	42	0	13747931	0
tiger:reviewed	25	13643979	420	1571	13641988	0
tiger:separated	39	11442206	4	0	11442202	0
tiger:name_base	914615	7422266	4	0	7422262	0
attribution	36	6767234	5470923	10169	1286142	0

The table on the home page lists the keys used in OpenStreetMap, along with the number of different values used for each key, and the number of times each one is used on nodes, relations, ways, and changesets. You can sort the table by clicking on any of the column headings. You can see the list of values used for any key by clicking on that key in the table.

You can filter the list of keys by typing text into the search box, which updates the table automatically. We need to find any possible tags to use for the club, so let's do that now. Type **amenity** into the search box, and the amenity key will appear along with several variations and misspellings. Click on **amenity** and wait for the table of values for this key to appear. This is similar to the table of keys, but with fewer columns. Again, you can sort the table by clicking on the headings.

We can filter the list of values in the same way as the keys, so enter **club** in the search box, and you'll get a list of values that match that string, as seen in the following screenshot:

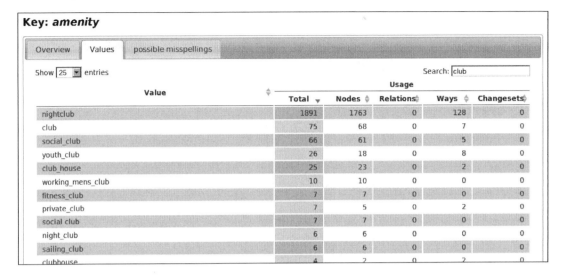

There are a few candidates we could use, but the most applicable appears to be `amenity=social_club`, so place a node for the club at this tag. Add the club's name in the `name=*` tag as well.

> **Contribute to the community by documenting tags**
>
> Normally, the first person to use a tag should write some brief documentation for it, but as with so many computer-related activities, this can be forgotten. If you find an undocumented tag like this one, you can help the OpenStreetMap community by creating a short page in the wiki, describing the tag.

Estimating area features

Just across the road from the club, in the corner formed by Spiceall and its spur, the trace forms a rough area shape. The photo in the middle of the area (`12112009107.jpg`) shows that this is a playground. The shape of the playground is rather irregular, and our trace doesn't give a good idea of the true shape.

When dealing with small area features like this, it's best to draw the shape of the feature in a notebook, but this isn't always possible. In some areas, high-resolution aerial imagery is available from Yahoo, which has said that tracing images is within its terms and conditions, and in some places from other sources. If it is, you can use this to map the shape of a feature.

Unfortunately, that's not the case for Compton, so draw a basic trapezoid shape for the playground. Tag it as a playground using the editor presets, as follows:

- In Potlatch, choose the football icon, then select **playground** from the drop-down list.
- In JOSM, choose **Travel | Leisure | Playground** from the **Presets** menu.
- Merkaartor doesn't have a preset for a playground, so tag it manually as `leisure=playground`.

The playground is surrounded by a fence, which we can map using a `barrier=fence` tag. Add this to the way in addition to the existing tag. A gate is located roughly half way along the eastern side of the playground, which we can map by adding a node to the way and tagging it with `barrier=gate`.

From the photo of the playground, we can see that it's provided by Compton Parish Council, and this information can be recorded using an `operator=*` tag. Again, this information may not seem very useful at first, but record any factual information like this that you come across.

One last area feature to map is the allotments in Compton. During the survey, it wasn't possible to gain access to the site itself. It's private land and the gates are locked, so we don't have a true representation of the size and shape of the site.

What we did record, however, is where the boundary of the site met the surrounding roads, and this is what the two waypoints named **Allotments1** and **Allotments2** represent. From this, we can make a rough approximation of the site by drawing straight lines from these points. You should end up with something like the following:

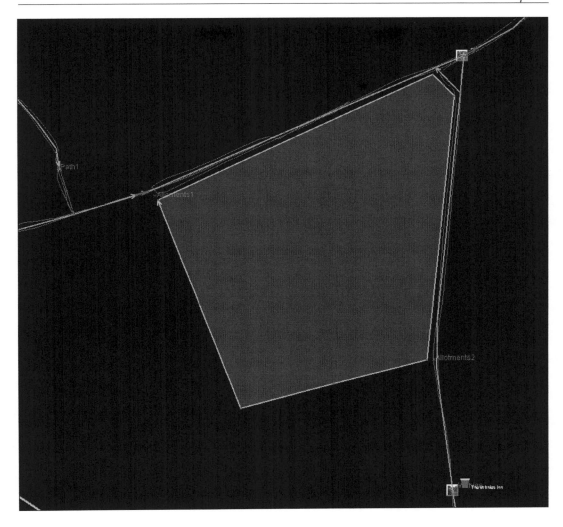

Tag the area as allotments using presets in Potlatch and JOSM, as follows:

- In Potlatch, choose the landuse icon, represented by three rectangles, then choose **allotments** from the drop-down list.

- In JOSM, choose **Geography | Landuse | Allotments** from the **Presets** menu.

- In Merkaartor, tag the area manually as `landuse=allotments`.

If you didn't map the remaining postboxes and pubs before, complete them now.

Finished

Congratulations! You've completed your first bit of mapping. Normally at this stage you would upload the mapping you've done to the OpenStreetMap server, describing the mapping you've done in the changeset comment field. After uploading the data, the two main renderings on openstreetmap.org should be updated with the changes after a short delay, which may be as short as a few minutes, but could be several hours if the load on OpenStreetMap's servers is high.

You can compare the mapping you've done with the real data for Compton to get an idea of how your work compares to that of other mappers. If you do download the live data for the area, you'll see more features than we've mapped here, including land-use areas, bus stops, some private residences, and the continuation of the roads that run through the village.

Summary

We've learned a lot in this chapter:

- How to draw features based on GPS traces
- How to tag them and how to find which tags to use:
    - Tagging presets in the editors
    - Documented tags in the wiki
- How to find tags that mappers have used, but not documented

Although we've used the most popular editing applications, others do exist, but the mapping techniques we've seen will apply to all OpenStreetMap editors to a large extent. They're the techniques and tools you'll use in whichever editor you use—even the ones that haven't been developed yet—and the methods won't change even if the precise tags used for any type of feature change.

7
Checking OpenStreetMap Data for Problems

You've surveyed, you've mapped and tagged, and you're nearly ready to use the data you and other mappers have collected. Just before you do, it's worth doing some checks to see if you or anyone else has made any mistakes. There are many online tools you can use to inspect the data and identify possible problems, provided by the project itself and some third parties.

In this chapter, we'll look at some of the following tools you can use to check OpenStreetMap data in a particular area, and what problems they can and can't tell you about:

- The data inspection tools on openstreetmap.org
- The NoName map
- ITOWorld OSM Mapper
- Geofabrik's OSM Inspector

Some of these tools identify specific problems, while others allow you to see when changes to the data were made and by whom. In either case, all these tools can do is offer guidance, not black-and-white answers, and you always need to use judgment when checking data.

It's important to remember that there are few fixed ideas of what is "wrong" data in OpenStreetMap. It should certainly be an accurate representation of the real world, but that's not something an automatic data-checking tool can detect. There may be typographical errors in tags that prevent them from being recognized, but there are also undocumented tags that may accurately describe a feature, yet be unknown to anyone except the mapper who used them. The latter is fine, but the former is a problem.

It's tempting to use the two map renderings on openstreetmap.org as a debugging tool, but this can be misleading. Not every possible feature is rendered, and many problems with the data, such as duplicate nodes or unjoined ways, won't be obvious from a rendered map. If a feature you've mapped doesn't render when a similarly tagged one does, there's an issue, but a feature appearing in the map doesn't mean it's free of problems, and a feature that doesn't appear isn't necessarily wrong.

Ultimately, you will have to use your own judgment to find out whether or not an issue reported by one of these tools is really an error in the data. You can always contact other members of the OpenStreetMap community using the tools we discussed in Chapter 2 if you're unsure what to do.

This is only a selection of the more widely used quality assurance tools used by mappers. For a more complete list, refer to `http://wiki.openstreetmap.org/wiki/Quality_Assurance`.

Inspecting data with openstreetmap.org's data overlay and browser

The openstreetmap.org website has a range of tools you can use to inspect the data in the database, both current and past. Some of the tools aren't obvious from the front page of the site, but are easily found if you know where they are.

The tools, which consist of the data map overlay and the data browser pages, allow you to see the details of any object in the OpenStreetMap database, including coordinates, tags, and editing history, without the need to launch an editor or read raw XML. As these tools work directly with the data in the OpenStreetMap database, they always show the most up-to-date information available. However, they simply provide raw information, and don't provide any guidance on whether the geometry or tagging of any feature could be problematic.

The easiest way of inspecting data is to start with the data map overlay. Go to the map view and find Compton (or any other area you want to inspect). Open the layer chooser by clicking on the **+** sign at the top-right. Click the checkbox labeled **Data**, and a box will appear to the left of the map view. After a short delay, the data overlay will appear, and a list of objects will appear in the box.

JavaScript speed and the OpenStreetMap data overlay

The data overlay and the accompanying list of objects make heavy use of JavaScript in your browser, and depending on how many objects are currently in your map view, can use a lot of processing power and memory. Some older browsers may struggle to even show the data overlay. Mappers have reported that Firefox 3.5, Apple Safari, and Google Chrome all work well with the data overlay, even with large numbers of objects.

Once the data for the area you're inspecting has loaded, you'll see something like the following image:

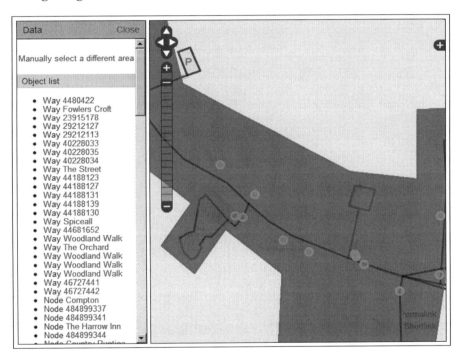

In the preceding image, on the left you can see the **Object list**, which gives a text description of every feature in the current map view, giving its primitive type and either its ID number or a name, if the feature has one. On the right is the map with the data overlay, which highlights every feature in the current area, whether they're rendered on the map or not. This last point is worth repeating: Not every type of feature gets rendered on the two map renderings used on openstreetmap.org, and those that do can take some time to appear if the load on the rendering engines is high. Any feature in the database will always appear in the data overlay.

Inspecting a single feature

To inspect an individual feature, either click on its entry in the object list, or on its highlight in the map view. Both the object list and the overlay will change to reflect this. Occasionally, an area feature may get drawn on top of other features, preventing you from selecting the ones underneath, but you'll still be able to select them from the list.

Let's select **The Street** and inspect its data. Either click on its name in the object list, or on the way in the map view, and the object list should change to show the tags applied to the feature, and you should see something like the following in the object list:

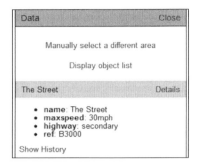

This gives a list of the tags attached to the feature. If you click on **Show History**, a list of the edits made to the current feature is added to the list. To get more information, click on the **Details** link next to the feature's name, and you'll be taken to the data browser page for that object, as follows:

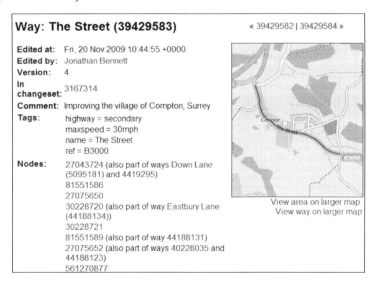

Here you see far more details about the feature we're inspecting. Apart from the object ID and its name, you can find the time when the object was last edited and by whom, and in which changeset. There are clickable links to any related objects and a map showing the feature's location.

At the bottom of the page are links to the raw XML of the feature, the history page of the feature, and a link to launch Potlatch — the online editor — for the area surrounding the feature.

Checking a feature's editing history

The OpenStreetMap database keeps every version of every feature created, so you can inspect previous versions and see when and how a feature has changed. To look at a feature's history, click on the link at the bottom of its data browser page. For the Watts Gallery in Compton, you should see something like the following:

You can see each version of the object listed in full, including which mapper created that version in which changeset, and what the tags for that version were. There's currently no way of showing any previous version or the changes between versions on the map, but third-party tools such as OSM Mapper provide some of these features.

Inspecting changesets

Along with looking at individual features, you can see how the map gets changed by looking at changesets. Since version 0.6 of the OpenStreetMap API went live in April 2009, every change to the map has to be part of a changeset. A changeset is a list of related edits made to OpenStreetMap data, with its own set of tags. What goes into a changeset is entirely up to the mapper creating it.

You can view the list of recent changesets by clicking on the **History** tab at the top of the map view. This will show a list of the 20 most recent changesets whose bounding box intersects your current map view. Note that this doesn't guarantee that any changesets listed include any edits in your current view, and any changesets covering a large area will be marked with **(big)** in the list.

Changesets

Changesets within -0.6761,51.1639,-0.5443,51.2454

« Previous | Showing page 1 | Next »

ID	Saved at	User	Comment	Area
#3597448	January 11, 2010 18:25	FK270673	international names	-180.000,-2.672,180.000,76.061 (big)
#3593999	January 11, 2010 08:54	TimSC	local knowledge	-0.587,51.242,-0.587,51.242
#3592981	January 11, 2010 00:45	sanchi	bridge	-122.421,-35.387,172.049,70.327 (big)
#3592926	January 11, 2010 00:30	sanchi	tunnel	-3.941,40.568,5.074,52.683 (big)
#3579922	January 09, 2010 17:36	TimSC	from memory	-0.605,51.240,-0.605,51.240
#3579734	January 09, 2010 17:19	TimSC	fix landuse	-0.621,51.224,-0.572,51.242
#3579144	January 09, 2010 16:24	Langlaeufer	User:Langläufer/Loipemap	-2.950,-45.648,160.497,70.325 (big)
#3578998	January 09, 2010 16:09	Langlaeufer	User:Langläufer/Loipemap	-4.268,-37.028,160.530,68.578 (big)
#3576321	January 09, 2010 11:42	Langlaeufer	User:Langläufer/Loipemap	-121.634,-38.341,160.612,73.281 (big)
#3574538	January 08, 2010 23:19	MichaelCollinson	typo is_in_continent --> is_in:continent	-6.959,49.381,22.853,54.586 (big)
#3572332	January 08, 2010 18:18	1248	Korrekturen	-1.393,51.000,8.537,51.963 (big)
#3571824	January 08, 2010 17:15	1248	Promenade ist KEIN designierter Radweg (kein blaues Schild), sondern Serviceweg mit erlaubter Fahrradnutzung	-1.395,50.999,7.636,51.969 (big)

Finding unsurveyed areas with the NoName layer

The NoName layer is a complete rendering of OpenStreetMap designed to highlight the urban areas that may need attention. To view the NoName layer, open the layer chooser in the map view on openstreetmap.org and choose **NoName** from the list. In NoName, any way tagged as a road but without a corresponding name=* tag is highlighted in red, as seen in the following image. The NoName layer was created to show areas where streets had been traced from aerial imagery, but no follow-up survey had been done to collect the street names.

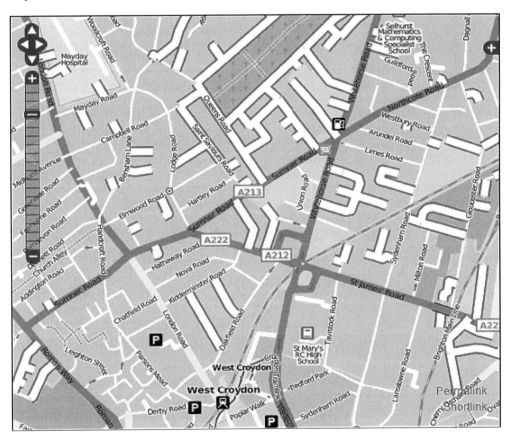

Large numbers of adjacent streets highlighted in the NoName layer probably means that the area needs surveying on the ground, to collect the street names and other points of interest. However, a road or street not having a name in OpenStreetMap isn't necessarily an error: The street may not actually have a name, or at least not one that can be identified by surveying it. You shouldn't tag a road with a blank name or some other incorrect value just to make it disappear from the NoName layer.

The NoName layer is provided by CloudMade, which also provides the layer as a downloadable map for Garmin devices at `http://downloads.cloudmade.com/`.

OSM Mapper

OSM Mapper (`http://www.itoworld.com/product/osm/`) is a web-based tool provided by a British transport intelligence firm, ITO, and provides similar facilities to those on the openstreetmap.org website, but allows you to sort and filter the data for a given area. At present, OSM Mapper only processes ways, not nodes or relations, but is still useful for finding sets of data with particular tags, and edited by particular mappers.

You need to register at the ITO website to use OSM Mapper, including supplying an e-mail address, and responding to a confirmation e-mail. You won't receive any further e-mail from ITO after registering. The data in OSM Mapper is usually updated once per day, so any edits you make won't show up here immediately.

Creating an area to analyze

To look at the data using OSM Mapper, you need to define one or more areas that interest you. The size of an area is limited to prevent overloading the system, although the limit of around 40 miles across is still quite large. OSM Mapper is most useful on much smaller areas than this, such as a single town. When you first register, you'll be shown an example area of Ipswich in the UK. To create your own area, click on the outline arrow to the right of the area chooser in the top right-hand corner of the page, and choose **Create a new area**. Use the slippy map that appears to choose an area, give it a name, and click **Save**.

Filtering and sorting data

Once you've defined an area, you can start to analyze its data. OSM Mapper shows all ways as outlines, which it color-codes according to the filters you create. To add a filter, choose from the options listed to the right of the map viewer: **Tags**, **Sessions**, or **Users**. These filter the current view according to the tags applied to the ways, by editing "session" or depending on which mapper last edited them respectively.

A session in OSM Mapper refers to a group of edits by a single mapper over a short period of time. OSM Mapper predates the use of changesets in OpenStreetMap, and there may or may not be a correlation between an OpenStreetMap changeset and an OSM Mapper session.

Click on **Tags**, and a table showing all keys that appear on ways in your current area will appear, complete with usage counts for each one. You can click on the headers in this table to sort the values by that column. If you now click on **highway** in the list of keys, you should see something like the following image:

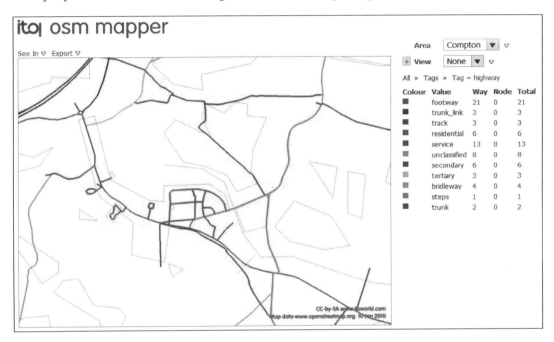

Any ways that don't have a highway=* tag have been faded out, and the remaining ways have been color-coded according to the value of highway=* used. The list has also changed to show those values and their usage counts. Clicking on a value will filter the view further to just the ways with that tag, and the list will change to show each one, listed by name if it has one, or ID if not. Click on an individual way and it will be highlighted, and its details will be shown in the list, including links that lead back to the data inspection tools on openstreetmap.org.

Creating a view

OSM Mapper allows you to create more complex combinations of filters than we've seen so far, and store them as Views so you can reuse them. Click on the **+** sign next to the **View** drop-down to the right of the map, and a box will appear. This lists the color table in use and any filters that apply to your current map view.

Let's create a view. Go back to the list of ways tagged as `highway=*`. To add a filter, use the arrow to the right of the current filter list. Click on this and you will get two filter options, to filter by the current tag being present or not present, so choose **Add: Tag Present: highway** to add this filter. You'll be taken back to the options, but you should now have a filter of **Tag present: highway** listed. Now click on **Users**, then **Jonathan Bennett**, and click on the filter drop-down and choose **Add: User=Jonathan Bennett**. You'll again be taken back to the filter options. If you click through to the list of tags again, you should now see something like the following screenshot:

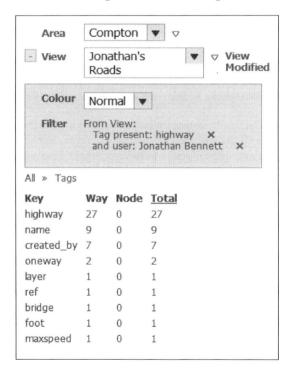

You can see the filters currently applied to the data, along with the table of usage numbers. These numbers represent the results of the filters, rather than the whole area. You can save the view by clicking the clear down-arrow to the right of the View name and choosing **Save the current view**.

You could add further filters to the view, such as a filter for roads with no `name=*` tag present, which would show you any street you may need to resurvey for a name, for example. Note that you can only filter to include a particular session, not exclude one.

OSM Inspector

OSM Inspector (`http://tools.geofabrik.de/osmi/`) is a web-based tool provided by German consultancy GeoFabrik, and it checks for several classes of potential problems with OpenStreetMap data. The interface looks as shown in the following screenshot:

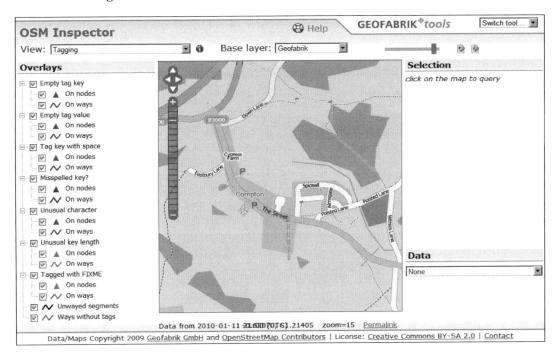

OSM Inspector doesn't require any registration or configuration; simply visit the site and use the slippy map to find the area in which you're interested. It's updated once every two days, so again your edits won't show up straight away. The date of the current data in use is shown below the slippy map. Across the top of the page are the view controls: a **View** drop-down list that chooses which class of problem to look at, the **Base layer** drop-down that allows you to choose which map rendering to use, and a transparency slider for the slippy map.

OSM Inspector has four Views that work for the whole world:

- **Geometry** tells you about problems such as self-intersecting ways, duplicate nodes in ways, and very long ways
- **Tagging** tells you about possible typographical errors in tags or other problematic keys and values
- **Places** identifies settlements and highlights, unnamed places, and problems with population tags
- **Highways** shows problems such as unnamed roads or ones without a reference number

The remaining four views are as follows:

- **Addresses** looks for problems in addressing information that has been added to buildings and streets
- **Boundaries** shows administrative boundaries such as state and county borders
- **Water** shows any type of waterway, and problems such as unnamed rivers
- Five **public transport**-related views show details of railways, bus routes, ferries, and other public transport infrastructure

There are also several country-specific views that have been used by mappers importing data in those countries; you won't need to use these.

Choose any one of these views, and a list of tests appears in the left-hand column. Each test has a checkbox, allowing you to turn it on and off in the slippy map, and a key for how it's represented on the map. Tests are also arranged in groups, and you can turn whole groups on and off using its checkbox. Hovering your mouse pointer over a test name will pop up a description of what that test checks for, and which zoom levels it shows up at.

Clicking on any highlighted feature in the slippy map will show its details in the right-hand column, as follows:

Rather than showing the tags applied to the selected object, this panel shows information relating to the problem OSM Inspector has identified with the feature. In the preceding screenshot, you can see that Compton is tagged as a village (`place=village`), but has no corresponding `population=*` tag. There are four icons near the top of the pane:

- The magnifying glass zooms the map to the selected feature
- The icon with a P launches Potlatch centered on the selected feature
- The icon with a J will center the current view in JOSM on the selected feature, provided JOSM is already running and has the Remote Control plugin installed
- The map and magnifying glass icon opens the openstreetmap.org data browser for the selected object

Below the **Selection** pane is the **Data** pane, which will show a list of all objects matching one of the tests in the current view. Click on the magnifying glass icon next to a feature in this list in to zoom to it.

OSM Inspector is very well documented in its own online help and in the OpenStreetMap wiki. Click on the **Help** link at the top of the page to access the online help.

Summary

Mistakes in data are inevitable, but that doesn't mean you can't correct them. It should be possible to find the cause of any problems you're having with OpenStreetMap data using one or more of the tools we've looked at:

- The data inspection tools on openstreetmap.org
- The NoName layer
- ITOWorld OSM Mapper
- Geofabrik's OSM Inspector

However, these tools can only provide guidance, and you should always use common sense when interpreting their results. Remember to ask the OpenStreetMap community if you're in any doubt about a possible error in the data.

8

Producing Customised Maps

Once the data of the area in which you're interested is complete enough for your purposes, you can produce maps to your own specifications. In this chapter, we'll cover how to produce one-off images of any area, including maps with custom styling.

In this chapter, you'll learn how to create maps:

- Using the standard renderings on openstreetmap.org
- Using a standalone rendering application for Windows, Kosmos
- In Scalable Vector Graphics (SVG) format using Osmarender

These maps will be static images that you can print out, send via e-mail, and include in a website or a printed publication.

The openstreetmap.org exporter

The openstreetmap.org website has a built-in map export function that you can use to produce images in various formats, at any scale using either of the example renderings on the site. The exporter doesn't allow you to change the style of the map.

To use the exporter, click on the **Export** tab at the top of the slippy map. A sidebar will appear to the left of the slippy map showing the exporter controls:

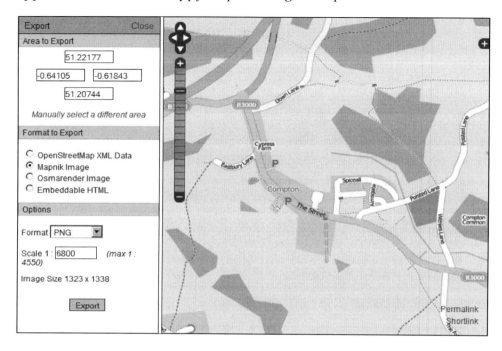

The default area to export is the whole of the map view. You can change this area by changing the map view itself, which should be reflected in the coordinates shown at the top of the sidebar. Alternatively, you can click **Manually select a different area** and drag a rectangle over the area you'd like a map of. Finally, you can enter the coordinates of the area you want to map directly into the form, and a rectangle will be drawn on the map to show the area you've specified.

If you need to save the location of an area you're exporting, you can use the Permalink in the map, but this won't save any adjusted coordinates, nor any options you've selected in the Export pane.

Choosing an image format

The exporter generates maps in four formats:

- OpenStreetMap XML Data
- Mapnik image
- Osmarender image
- Embeddable HTML

You use the first option to produce data to use with a separate rendering application, such as Kosmos or Osmarender, covered later in this chapter. The data returned will include entire features, including parts that lie outside the area you've specified, but we'll see how to trim this down later in the chapter.

The second and third options produce image files using the same rendering technologies as the main slippy map, and the fourth allows you to link to the exporter directly from a web page, which keeps the image up-to-date without further action.

Generating image files

The Mapnik and Osmarender image formats create a single image file of the area you've chosen, using a variety of file formats. Both support:

- Portable Network Graphics (PNG)
- JPEG

In addition, you can export Mapnik maps in these following formats:

- Scalable Vector Graphics (SVG)
- Portable Document Format (PDF)
- Postscript

Note that the SVG images produced by the exporter are bitmaps, not vector graphics of the map. We'll use Osmarender to generate true vector maps later in the chapter.

Let's export a Mapnik image in PNG format, so select those two options. You should see something like the following screenshot at the bottom of the exporter pane:

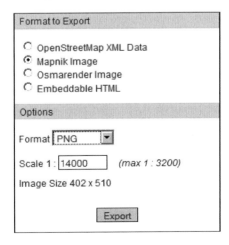

The last option to complete is the scale we want our map to have. For a Mapnik image, this is specified as a ratio, in common with most printed maps. The higher the number, the larger the scale, and the smaller the image. The ratio has to be lower than the maximum given, so the number you type into the box must be larger than the one shown to the right. If you enter too large a ratio, the **Export** button will be disabled.

Below the scale box, you're given the size in pixels of the image the exporter will produce. The size is dependent on both the area you've selected and the scale you've chosen. Adjust the scale until you get the image size you want, remembering you can edit the image after export in an image editing program to trim it to an exact size.

The Osmarender export works in a slightly different way. Instead of choosing a scale, you pick the zoom level to use from the slippy map. You're not given the exact image size that will be produced in advance, but for each increase in zoom level, the image's width and height in pixels will double, increasing the image area by a factor of four.

Once you have the area you want, at the size you want, click **Export**, and your map will be generated. This could be almost instant, or have a short delay depending on the load on the servers, and under some circumstances, the service may be temporarily suspended by OpenStreetMap's system administrators to keep the load on the servers down.

Embedding maps in a web page

The final exporter option is to provide an HTML link that you can use to embed a slippy map in any web page that will be updated as mappers add data to OpenStreetMap. This is a good option if you want a map of an area that currently lacks data, but for well-mapped areas, it's a less reliable service than creating and hosting your own map.

Exporter service levels

The exporter service on openstreetmap.org, which provides the embedded maps, is provided on a "best effort" basis, and there is no guarantee that the service will be available at any time, and could be unavailable for an extended period or even permanently. If this is the case, your embedded map will not appear on your site. If this isn't acceptable to you, use one of the other options for creating a map.

Once you've chosen the area of which you need a map, you can choose to add a marker to the embedded map at a particular location. To do this, click on **Add a marker to the map**, and click where you'd like to add the marker. You can zoom in and out of the map without affecting the position of the marker, so you can place it precisely, then zoom out to get a larger scale map. In the following image, we've placed a marker on Compton Village Hall, which isn't shown in the Mapnik rendering:

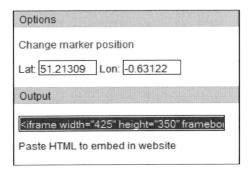

Once you've chosen your map area and any marker location, copy the generated HTML into the web page where you want your map. Consider the following HTML code:

```html
<html>
  <head><title>OSM Export test - Compton</title></head>
  <body>
    <h1>OpenStreetMap map of Compton, Surrey</h1>
    <iframe width="425" height="350" frameborder="0" scrolling="no"
      marginheight="0" marginwidth="0"
src="http://www.openstreetmap.org/export/embed.html?bbox=-
0.64138,51.20954,-0.61958,51.22107&layer=mapnik&marker=51.21303,-
0.63091" style="border: 1px solid black">
    </iframe>
    <br />
    <small>
<a href="http://www.openstreetmap.org/?lat=51.215305&lon=-0.63048&zoom
=15&layers=B000FTFTT&mlat=51.21303&mlon=-0.63091">View Larger Map</a>
    </small>
  </body>
</html>
```

This HTML code will give you a page similar to the following screenshot:

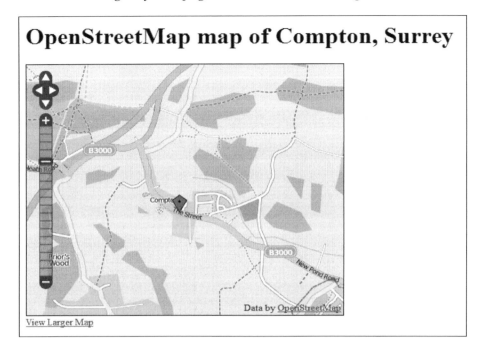

This is a fully working slippy map like the one on the openstreetmap.org home page, except you don't get the layer chooser or permalinks. You can edit the HTML to suit your site's design better, provided you don't change the src attribute of the iframe element.

Rendering maps on Windows using Kosmos

Kosmos is a .NET application that renders maps from OpenStreetMap data based on sets of rendering rules normally stored in the OpenStreetMap wiki. Kosmos was written by Igor Brejc, and is available in graphical and command line versions. It can render single images or small sets of tiles, and can even provide a small web server for those tiles.

Kosmos is designed for rendering a small number of features. It reads OpenStreetMap data directly from XML files, and doesn't have an internal database. If you want to generate a map of an area larger than a small town, Kosmos probably isn't the tool to use. Kosmos can produce less complex maps of larger areas, but this is best accomplished using data that only contains the features you want to show. We'll show you how to do this using the extended API (XAPI) later in the book.

Kosmos doesn't include any data editing facilities, so if you want to change any data from OpenStreetMap before rendering it, you'll need to edit it using JOSM or Merkaartor before loading it into Kosmos.

Kosmos also uses an unusual way of storing rendering rules for its maps: It normally retrieves them from the OpenStreetMap wiki, and uses some of the templates installed on the wiki to make writing rendering rules faster and easier. Once you've written the rendering rules you want, you can then store the generated rules file anywhere, including locally, but the easiest way of authoring the rules is online.

Development on the current version of Kosmos has finished, and a new program called Maperitive is being developed with a different interface and syntax for rendering rules. This means the few issues Kosmos has are unlikely to be fixed, but the program still works and you can render maps with it.

We're going to use Kosmos to create a map showing where Compton Village Club is, suitable for use in a brochure or on a website.

Installing Kosmos

To install Kosmos, you need a platform that supports .NET 3.5 SP1 or higher, although the latest version of .NET is recommended. This means Windows XP or later is supported, although there are efforts to make Kosmos run on the Mono platform on Linux.

You can download the latest version of the .NET framework from `http://msdn.microsoft.com/netframework/default.aspx`.

You can download the latest version of Kosmos from `http://downloads.igorbrejc.net/osm/kosmos/Kosmos-latest.zip`, and install it by unzipping the archive into a folder. You'll need to create shortcuts on your desktop or Start menu if you want either of these.

Creating a project

Once you have Kosmos installed, launch it by opening the folder into which you unpacked the archive, opening the **Gui** subfolder and double-clicking on **Kosmos. Gui.exe**. You should see the welcome screen shown in the following screenshot:

Kosmos has a simple user interface that reflects its role as purely a rendering application. All work in Kosmos is done as part of a project, so let's create a new project by clicking on **Start with an empty Kosmos project** on the welcome page. Kosmos should now show its default workspace, with a map taking up most of the workspace, the **Project Explorer** top-left, the **Properties** pane bottom-left, and the **Activity Logger** and **Error List** stacked across the bottom of the map space. All the panes in the interface are re-sizable or can be hidden. Click on the pushpin icon to hide a pane.

For an empty project, the **Project Explorer** will look as shown in the following screenshot:

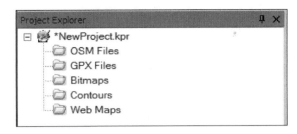

You can see the types of files you can add to a Kosmos project, including:

- OpenStreetMap data
- GPX traces
- Contours
- Existing web maps

By using different types of data and images, you can illustrate many kinds of geographical information without necessarily having to add the data to OpenStreetMap.

Adding OpenStreetMap data

To load OpenStreetMap data into Kosmos, you have two options:

- From a local XML file
- Via the Extended API (XAPI)

The latter of these options uses a third-party XAPI server, rather than the main API at openstreetmap.org, and can be unreliable as well as being unavailable offline. However, it does mean having to use a separate tool to obtain the data.

We're going to add data from a local XML file. You can create this yourself using the OpenStreetMap exporter, or use `compton.osm` from the download file for this chapter. Right-click on **OSM Files** in the **Project Explorer** and choose **Add File(s)** from the pop-up menu. Choose the file, and click **Open**.

After the data has loaded, the map will be rendered using Kosmos' default rendering rules. If you don't see anything after the data has loaded, right-click on the data file in the **Project Explorer**, and choose **Show on Map**. This zooms the current view to the bounding box of that file, and you can do this with any data or GPX file.

Now that we have some data loaded, the **Properties** pane shows some information:

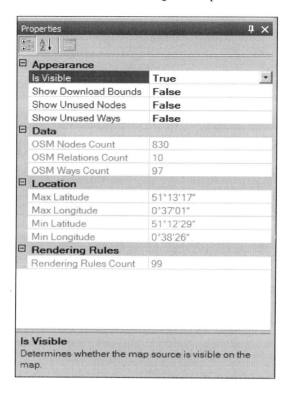

Most of the properties for a data file are read-only, but there are some properties that affect the map rendering. You can change the visibility of any file used in your map project, so you can produce variations on the same map without needing separate projects for each one.

Kosmos can also show any nodes or ways that don't match any of the rendering rules in the current ruleset by setting **Show Unused Nodes** and **Show Unused Ways** to **True**. This is a useful debugging tool, as it allows you to see whether something you're expecting to see on the map is actually in your current data. To see the details of any feature in the map, including ones highlighted as unused, right-click on the feature and choose **Show Object Info** from the pop-up menu. A pop-up will appear showing the ID and tags for the feature.

Adding GPS tracks

Although Kosmos isn't an editing application, you can still use GPS traces in GPX format as part of your map. This is a quick way of showing a route on a map, provided you have the data in the right format.

To add a GPS trace, right-click on the **GPX Files** folder in the **Project Explorer** and choose **Add File(s)** from the pop-up menu. Kosmos can also use GPSbabel to get traces directly from your GPS receiver if you have it installed, but this isn't installed here.

You can't change the rendering of GPS tracks in Kosmos, so any GPX files you add to your project will be drawn as thin red lines with squares for trackpoints at higher zoom levels, and named waypoints in blue. You can show or hide tracks, trackpoints, or waypoints separately.

Adding existing OpenStreetMap maps

Along with the map you're rendering from the data you have, you can use the map tiles that have already been rendered as a background. This can simplify generating a map, as you only have to render any feature you want to highlight, rather than the entire map.

To add a map from the openstreetmap.org website, right-click on **Web Maps** in the **Project Explorer**, and choose **Add Web Map** from the pop-up menu. Choose one of the renderings to use. The **Draw in background** option makes the map tiles appear behind all other features, and for most purposes, you should tick this box. You can choose the transparency level of the background, and whether to download the tiles from the server or use a local cache; for most purposes, downloading the tiles is the best option.

You can adjust any of the options for a web map after creation by selecting it in the project explorer and editing the properties.

Customizing the rendering rules

So far, we've just seen the map rendered using the default rendering rules. Now we can customize these rules, so the map looks the way we want it to look. To do this, we need to write our own set of rendering rules, and this is normally done by creating a page in the OpenStreetMap wiki.

To make authoring rules for Kosmos easier, a number of the standard templates on the wiki are used. It also means that the rules are formatted nicely, and so easier to read when you view the wiki page. Each rendering rule is a single line in a table.

To get started, it's easiest to copy an existing page of rendering rules for Kosmos and strip out the content you don't want. You'll need an account on the OpenStreetMap wiki, so if you didn't create one earlier, go to `http://wiki.openstreetmap.org/` and register.

Go to `http://wiki.openstreetmap.org/wiki/User:Jonobennett/Compton_Rendering_Rules` or `http://bit.ly/kompton`, and click on **View Source** or **Edit**, depending on whether you're logged into the wiki. Now, copy the contents of the edit box for the page into a new wiki page below your own User page. To do this, click on your wiki username at the top, and once the page has loaded, add `/Compton_Rendering_Rules` to the end of the URL, hit return, and then click on **edit this page**. Now paste the rules into this page, and click on the **Save page** button at the bottom of the page.

To make Kosmos use this page, click on the project in the **Project Explorer**. The **Properties** box should then show a single item—**Rendering Rules Source**. Paste the URL of your rules page into the box. Now press *Control + R* to reload the rendering rules, and the appearance of the map should change:

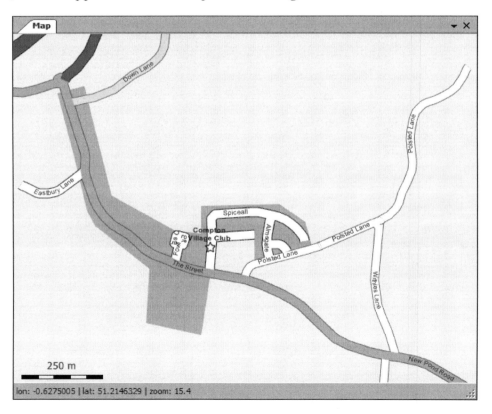

Rendering rules not reloading

There is an issue in Kosmos that means rendering rules don't always reload when you request it, meaning any edits you make to the rules won't show up in the map. You can work around this by restarting Kosmos, or by copying the rules to a new page and changing the rules source.

The page has two sections: the options and the rendering rules. The options specify the minimum version of Kosmos that will understand these rules, which is set to 2.1 in our page, but isn't important as we're using a later version than this. Two other options, `LandBackgroundColor` and `SeaColor`, set the colors used for areas not otherwise by a rendering rule. You can copy this options section to any page of rules you create, and edit the colors if necessary.

A Kosmos rendering rule is a single line in a table with the following structure:

```
| <title> || <targets> || <tag selector> || <template> (<parameters>)
|| <rendering options> || <comment>
```

The vertical bars are part of the wiki formatting, separating the cells in the table. If any rule you create skips a column, you still need to enter the bars with whitespace between them to make the table display properly in the wiki. Each rule is separated with a table row separator:

```
|-
```

The items in each rule are:

- The **title** is a unique name for the rule. If a title begins with a period, it's classed as a "child" rule. Child rules modify the parent (non-child) rule directly above them in the table.

- **Targets** use a standard template to identify to which OpenStreetMap primitive type this rule applies. The exact format is one or more of the following:
    - `{{IconWay}}`
    - `{{IconNode}}`
    - `{{IconArea}}`
    - `{{IconRelation}}`

    You can use more than one target per rule, but this isn't done very often.

- The **tag selector** uses another template to choose only features with certain tags. The tag template uses either `{{tag|<key>}}` or `{{tag|<key>|<value>}}`. You can use multiple templates to match the features with every tag you list.

- The **template** and **rendering instructions** define how the features selected by the rule are shown on the map. We'll cover these in more detail shortly.

- The **rendering options** change the behavior of the renderer for that feature. Only one option, EliminateSeams, is available. This smoothes out the edges of adjacent features of the same type, and only applies to polyline features.

The template tells Kosmos what shape to use to render matching features. The available templates are:

- Polygon for closed areas
- Polyline for linear features
- Text for labels
- Symbol for icons drawn on a linear feature, such as route numbers
- Icon to draw an icon for a point feature

For each template, there are a number of parameters you add to specify how a feature is drawn. They're all documented on the OpenStreetMap wiki at http://wiki.openstreetmap.org/wiki/Kosmos_Rendering_Help, so we'll just cover what we need for our map here. The general syntax is that the parameters are listed after the template in brackets.

The first feature we're going to render is the built-up area of Compton itself. These areas are normally mapped in OpenStreetMap as a closed way tagged with landuse=residential, so we need a rule that selects this:

```
| Residential     || {{IconArea}}     || {{tag|landuse|residential}}
```

Let's render this as a pink-shaded area. This is a closed area, so we use the polygon template. To give this a color, we use the Color option, which takes an HTML color as an argument, either a color name or an RGB value in hex. We'll use:

```
color=#FF9999
```

That's the only option we're going to use, so our rule looks as follows:

```
| Residential || {{IconArea}} || {{tag|landuse|residential}} ||
Polygon (Color=#FFBBBB) ||   ||
```

If we wanted a border around the area, we could use the BorderColor, BorderWidth, and BorderDashStyle parameters.

Next, we need to add the roads to the village. The trunk A3 road passes just to the north of the village and is included in our data, so let's render this as a green line with a black border. To select trunk roads, we need the following to select ways tagged as `highway=trunk`:

```
| HighwayTrunk || {{IconWay}} || {{tag|highway|trunk}}
```

To render this, we use the `polyline` template. Again, this takes a `color` parameter, but we also want to add a border, so we use the `BorderColor` parameter:

```
bordercolor=black
```

We also need to specify the width at which we want to draw the roads, using the `width` parameter. The value for this parameter takes a complex value based on the width we want the feature to have at varying zoom levels. For each zoom level you specify, you give a size for the feature, separated by a colon. You can give values for multiple zoom levels separated by a semicolon. Kosmos will interpolate the value between the specified zoom levels:

```
Width=6:1;17:14
```

Finally, we use the `EliminateSeams` option to ensure that any roads drawn by our rule have smooth edges. Our final rule looks as follows:

```
| HighwayTrunk || {{IconWay}} || {{tag|highway|trunk}} || Polyline
(MinZoom=6, Color=#008000, BorderColor=black, Width=6:1;17:14) ||
EliminateSeams ||
```

We'll repeat this rule to render the link roads to A3, substituting `{{tag|highway|trunk_link}}` as the selector.

Other roads in the area are rendered using similar rules, except that we also want to add the names of these roads to the map in addition to the rules to draw the roads. For The Street, which is a secondary road in UK terms, we need a rule that selects these roads:

```
| HighwaySecondaryText || {{IconWay}} || {{tag|highway|secondary}}
```

To render the street name, we use the `text` template. This can produce text for any type of feature that has a suitable tag. We're also going to use an option that can be applied to any template—`MinZoom`. This option, along with its counterpart `MaxZoom`, limits when a feature is visible. This helps make maps more readable by hiding details, such as street names at lower zoom levels, or settlement names at higher zoom levels. We're producing a single static map at a fixed zoom level, so these parameters are less important, but if you were using Kosmos to produce a set of tiles for use in a slippy map or on a mobile device, you'd need to use them carefully.

You select which tag to use with the `TagToUse` parameter and the key of the tag, so we'll use the following:

```
TagToUse=name
```

We could also use `ref=*` to show the route code for a road. We also need to format the text using the `FontName`, `FontStyle`, and `FontSize` parameters, so our formatting instruction becomes:

```
Text (MinZoom=15, Color=black, TagToUse=name, FontName=Arial,
FontStyle=regular, FontSize=15:6;17:9)
```

`FontSize` takes values in the same format as `Width`, again with the size interpolated between explicit values. Specifying a `Width` or `FontSize` for a feature at a lower zoom level than any `MinZoom` parameter will result in an error.

We'll repeat similar rules for unclassified and residential roads, varying the color of the roads and making them slightly narrower.

```
| HighwayUnclassified || {{IconWay}} || {{tag|highway|unclassi
fied}} || Polyline (MinZoom=6, Color=white, BorderColor=black,
Width=6:1;17:14) || EliminateSeams ||
| HighwayResidential || {{IconWay}} || {{tag|highway|residential}} ||
Polyline (MinZoom=6, Color=white, BorderColor=black, Width=6:1;17:14)
|| EliminateSeams ||
```

The last detail to include on our map is the club itself. We need to show its location and name, so we'll use an `icon` template and a `text` template. We'll use the same selection for both rules, which will select a node tagged with `amenity=social_club`:

```
| SocialClub || {{IconNode}} || {{tag|amenity|social_club}}
```

We'll use a simple star to mark the location of the club, stored in the OpenStreetMap wiki at `http://wiki.openstreetmap.org/images/2/2c/Star-icon.png`. To display this, we use the icon template, which takes an `IconUrl` parameter, as well as `Width` and `MinZoom` parameters:

```
Icon (MinZoom=6, Ico
  nUrl=http://wiki.openstreetmap.org/images/2/2c/Star-icon.png,
  Width=6:8;17:16)
```

We'll also add the name using a `Text` template. We need to move the text upward so that it sits above the icon for the club, so we use the `TextLineOffset` parameter that shifts the text vertically, upward for positive values, and downward for negative values. The value itself is expressed as a percentage of the current text size:

```
Text (MinZoom=6, Color=black, TagToUse=name, FontName=Arial,
  FontStyle=Bold, FontSize=6:2;15:6;17:12, TextLineOffset=80%)
```

This is enough to show how to find the club, so we won't render any more details. You can see all the features we haven't rendered by selecting **compton.osm** in the **Project Explorer** and setting **Show Unused Ways** in the **Properties** pane to **True**.

However, the map looks odd, because the unclassified and residential streets are rendered on top of the secondary road through the village:

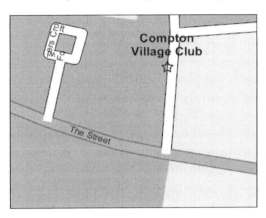

This is easily solved by changing the order of the rendering rules in the page, so the rules for secondary roads come last. If you had a more complicated set of rules, or wanted to keep your rules in order for readability, you could add
`Options=TopLayer`
to the rendering parameters to make that rule render after others in the same group. For a full description of how Kosmos layers different types of features, refer to `http://wiki.openstreetmap.org/wiki/Kosmos_Layering`.

Exporting a bitmap

Once you're happy with the rendering rules you have, you can produce a static image of the map. Kosmos can produce images in Windows Bitmap (`.bmp`), JPEG (`.jpg`), Portable Network Graphics (`.png`), and TIFF (`.tif`) formats.

To produce a bitmap, first get the view you want to use as your bitmap in the map viewer. If you can't get the precise size and shape of the image you want with the zoom level you want, you can crop the image afterwards in a graphics editing package.

Next, choose **Export to Bitmap** from the **File** menu, or press *Ctrl + E*, and the Export dialog box will appear:

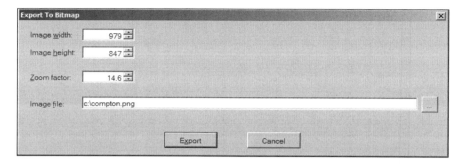

Here you have a chance to alter the zoom factor used for your image. Changing this will alter the size of the image, but not the area included, which remains the same. This allows you to produce larger images of areas at high zoom levels. To choose which format your export is in, you need to add the appropriate three-letter extension to your filename.

It is also possible to print directly from Kosmos, but unfortunately the print function has a bug where you can't set print area reliably, so you'll get better results by exporting a bitmap and printing in another application.

Kosmos Console

Kosmos also has a command-line version that can generate images from existing Kosmos project files. This has the advantage that you can use it in batch scripts or other automated operations. Along with generating bitmaps like those we've produced in the GUI version of Kosmos, the Console version can also generate sets of tiles for use in a slippy map or a mobile device.

To use Kosmos Console, you must have created a project file using the GUI version beforehand. This will specify the data and rendering rules used to produce the maps.

Kosmos Console takes a command line in following form:

```
Kosmos.Console.exe <command> <options>
```

To produce a single bitmap, use the `bitmapgen` command. This has the following form:

```
Kosmos.Console.exe bitmapgen <project file> <image file> <options>
```

The options for this command are the area and zoom of the map to generate. You can specify the area as a bounding box using latitudes and longitudes, the center of the image, the size of the image in pixels, or the zoom factor of the map. Kosmos Console needs two of these parameters to produce a bitmap. For instance, the following command will generate the map we've just designed:

```
Kosmos.Console.exe bitmapgen compton.kpr compton.png  -mb 51.20868 -
0.63794 51.22011 -0.61772 -z 16
```

Here we specify the minimum latitude, minimum longitude, maximum latitude, and maximum longitude in degrees, along with the zoom factor we want. This produces an image 942 x 850 pixels in size.

Rendering map tiles using Kosmos

You can also generate a set of map tiles using the `tilegen` command. These tiles can then be used with any number of mobile applications, or a slippy map like the one on openstreetmap.org. As with a single image, Kosmos isn't a good tool to use for generating map tiles covering a very large area, but for relatively small areas (a town or district), you can get good results a lot easier than setting up a full tile rendering system.

To generate a set of tiles for Compton, use the following command:

```
Kosmos.Console.exe tilegen compton.kpr 51.20868 -0.63794 51.22011 -
0.61772 15 18 -ts Tiles
```

This uses our project file, along with a bounding box, minimum and maximum zoom levels, and a target directory. When you run this command, a set of directories is created below the target using a system recognized by OpenLayers—the software used to provide the Slippy Map on openstreetmap.org, but which is also used by other map viewing applications. Note that because the tiles are of a fixed size, you won't get the bounding box you specify precisely, but rather the set of tiles needed to render your bounding box completely. If you don't have a complete set of data for this area, which is the case for our data file, you will be able to see this in the generated tiles.

Once your tiles have been created, Kosmos Console can provide a simple web server with OpenLayers to allow you to inspect your tiles. To start the web server, start Kosmos with the `tileserv` command:

```
Kosmos.Console.exe tileserv Tiles http://localhost/Kosmos/
```

The first parameter for this command is the directory where your tiles can be found, and the second, the URL that you want to use for your map. Once this is running, point your browser at `http://localhost/Kosmos/`, where you'll see a full-page slippy map showing lots of "broken" tiles. This is because the map view defaults to the whole world, and we haven't generated the tiles for this. Click the **+** sign at the top-right to open the layer chooser and switch to the **tiles@home (direct)** layer, which should allow you to see what you're doing. You can either zoom into Compton yourself and switch back to the **Kosmos** layer once you're there, or `http://bit.ly/localcomptonmap` should take you straight there. You should see two tiles covering the whole village:

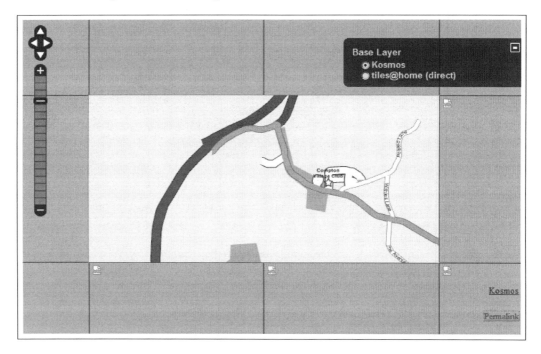

You can move around our mini-map in the same way as the main map on openstreetmap.org. If we used a data file covering a larger area, we could render more tiles, and at lower zoom levels. You can also set up your own slippy map using OpenLayers, and set the default view to where you've rendered tiles, but this requires JavaScript programming skills.

Taking Kosmos further

Although development on Kosmos has ended, and the bugs we've mentioned are unlikely to get fixed, it's still a useful tool for rendering maps of smaller areas. We've also only produced a relatively simple example here, but Kosmos is capable of rendering more detail than we used. If you want to see what other mappers have created using Kosmos, many of them have created rules in the OpenStreeMap wiki, listed in a category at `http://wiki.openstreetmap.org/wiki/Category:Kosmos_rules` or `http://bit.ly/kosmosrules`.

Osmarender

Osmarender is a custom-written rendering tool for OpenStreetMap data that produces Scalable Vector Graphics (SVG) output suitable for printing, using in a web page, or editing and converting to another format. It's the technology used to generate one of the example renderings on openstreetmap.org, but it's also well-suited to producing individual maps from a file containing OpenStreetMap data. The advantage of using SVG for a map is that you can choose the size of the image you want to use without losing image quality.

Osmarender isn't a traditional program, but an Extensible Stylesheet Language Transform (XSLT) document that an XSL processor uses to turn an XML document containing OpenStreetMap data into an SVG file according to a set of user-defined rules.

This means that Osmarender can be difficult to use at first, as it's driven from a command line interface, and you need to edit XML files to customize the rendering. You don't need a full understanding of the details of XML or XSLT to use Osmarender, but if you're not comfortable editing XML files, then this isn't the tool to use.

You'll also need a working knowledge of Cascading Style Sheets (CSS) to customize Osmarender maps, as this is what's used to add formatting and presentational information to the map. We'll show you some examples of how this works here, but there are more sample stylesheets available for Osmarender.

Getting ready to run Osmarender

There are a few prerequisites for running Osmarender:

- an XSL processor program
- the Osmarender files
- An SVG editing package or renderer

XSL processing with XMLStarlet

There are many XSL processors available for different platforms. In our example we're going to use **XMLStarlet**—an open source command-line XML utility that runs on Unix-like operating systems and Windows. You can use any program that performs a similar function.

On Linux, you will probably be able to install XMLStarlet using your package manager. If you can't find a package, or are on Mac OS X, the source, and for Windows, a binary distribution, are available from `http://sourceforge.net/projects/xmlstar/files/`.

To install the Windows version of XMLStarlet, unpack the ZIP file into a folder on your hard drive, such as `C:\Program Files\xmlstarlet`, and add the folder to your path. You can do this by right-clicking on your **My Computer** icon and selecting **Properties**. In the **Advanced** tab of the **System Properties** dialog (seen in the following screenshot), click on the **Environment Variables** button.

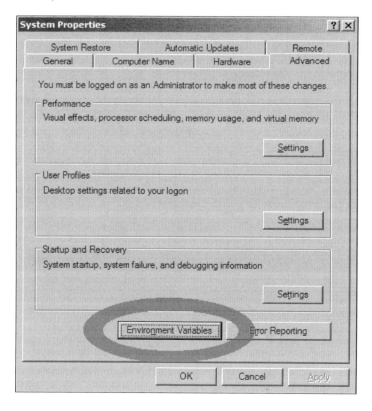

The **Environment Variables** dialog should appear. In the lower box titled **System variables**, find the **Path** variable and click on the **Edit** button. Add a semicolon and the path of the folder you installed to the value (;C:\Program Files\xmlstarlet in our case). Click **OK** on each open dialog until they're all closed.

You should now have XMLStarlet installed. To check if this has worked, open a command-line window, and try the following command on Unix-like operating systems:

```
xmlstarlet --version
```

Try the following command on Windows:

```
xml --version
```

Note the difference in syntax between Unix/Linux and Windows. Our examples will use the former of these commands. If XMLStarlet is properly installed, you should see the version number returned, which was **1.0.1** at the time of writing.

Installing Osmarender

As Osmarender isn't a normal application that you install, there's no simple download of Osmarender available, and you may have to download several files.

 If you're familiar with the version control system Subversion, you can check out the whole of Osmarender from http://svn.openstreetmap.org/applications/rendering/osmarender/, including the style sheets used to create the Osmarender layer on openstreetmap.org. If you do this, the main Osmarender XSLT file is in the xslt subdirectory. For our examples, you'll need to copy the file osm-map-features-z17.xml from the stylesheets subdirectory into xslt. You can then use xslt as your working directory.

To get a minimal working Osmarender installation, download the following files to the same folder:

- http://svn.openstreetmap.org/applications/rendering/osmarender/xslt/osmarender.xsl

- http://svn.openstreetmap.org/applications/rendering/osmarender/stylesheets/osm-map-features-z17.xml

The first of the two preceding files is the Osmarender transforms file that does all the work, and the second is the rules file used to generate one zoom level of the Osmarender layer on openstreetmap.org. You can explore the other files that are used in conjunction with Osmarender, but for our examples, those two are all you need.

Editing SVG with Inkscape

Finally, it's useful to have a vector graphics editing package that understands SVG, so that you can make final alterations to your map before publication, and produce bitmap versions if necessary. If you don't already have such a package, you can download Inkscape (`http://www.inkscape.org/`) for free. Inkscape is an open source, cross-platform vector editing application that uses SVG as its native format. It runs on Windows, Mac OS X, and many Unix-like operating systems.

If you're running Linux, you'll probably be able to install Inkscape using your distribution's package management utility. If you're running Windows or Mac OS X, you'll need to download the installer from the Inkscape website and run that.

Producing a map

Now that we have all the software we need, we can produce a map. We need a set of data for the map we want to produce, so let's use Compton as our example again.

Copy the file `Compton.osm` to the same folder to which you downloaded the Osmarender files. The standard rules files for Osmarender take their data from a file named `data.osm`, so rename your data file to this.

To generate the map, open a command-line window, change to our working folder, and use the following command:

```
xmlstarlet tr osmarender.xsl osm-map-features-z17.xml > compton.svg
```

This command tells XMLStarlet to use the transforms in `osmarender.xsl`, apply them to `osm-map-features-z17.xml`, and store the results in a file named `compton.svg`.

Osmarender will print out the rules it's processing as it goes along. If you have only downloaded the Osmarender file and a standard stylesheet to make this work, you'll see warnings about external entities not loading. This means that some symbol files created for the standard rendering won't get used, but the map should still be generated without the missing symbols.

Once the processing is finished, you should have a file containing an SVG map of Compton. You should be able to view this using any web browser except Internet Explorer. However, it will appear very small by default, so you may have to zoom in to see the details. What you will notice is that the map is clipped to the bounding box of our download, unlike in Kosmos. Despite there being data on our file outside this area, Osmarender can read the bounding box in the data file, and never displays any features outside this area.

It's easier to see the map if you open it in an editing application. The following is the map in Inkscape:

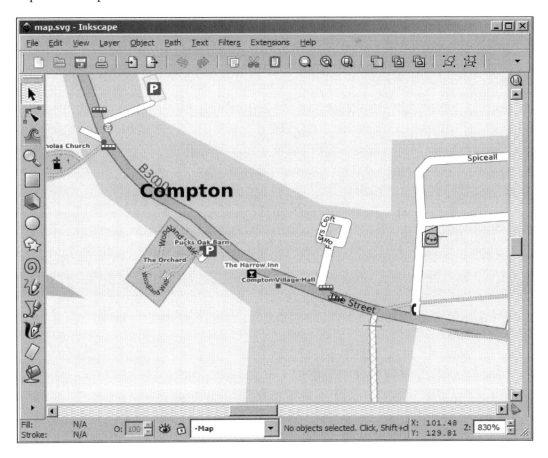

Your version may be missing the symbols for some of the points of interest, but the names should still appear as captions.

We have a map, but that's not the end of the story. This map is a fully editable document, so we can still alter it without losing any quality. For instance, in the previous image, the caption for the village itself sits over The Street, due to the placement of the node representing the village. We can now move that text in our editor so that it doesn't overlap any other features. We can change the color, size, or font of any feature or caption as we see fit.

We can also add anything we like to the map, so if we want to mark a single location on the map, we can do it at this editing stage, rather than having to write a custom rendering rule just for that feature. We can also delete any feature we don't want to show on our map without affecting the data. Essentially, the only limit to our ability to change this map is what we're able to do with our editing application.

Finally, we can produce a bitmap version of our map once all our alterations are complete, to any size we like and in any format our editing application supports. To do this in Inkscape, go to the **File** menu, and choose **Export**.

Customizing the rendering rules

We've got some fantastic results so far—a scalable, editable map of Compton—without having to edit any XML files by hand. However, if we want to use the full power of Osmarender, we need to create our own rendering rules.

As we've already seen, a map is produced using an XML rules file. This contains all the instructions about how any features should be displayed on the finished map.

In its simplest form, a rendering rule for Osmarender is a single rule element with a rendering instruction:

```
<rule e="way" k="highway" v="primary">
  <line style="stroke-width: 1px; stroke: black" />
</rule>
```

This rule selects any way tagged as highway=primary and renders it as a 1-pixel-wide black line. Let's have a look at how that's done:

- In the rule definition, the e attribute can be either node, way, or relation to render those elements respectively. You can also match multiple primitive types by separating them with the bar character |, so a match for both nodes and ways would use e="node|way". Many points of interest can be mapped using either primitive, so for rules designed to match these features, it's best to match both.

- The k and v attributes specify the key and value of the tag that this rule should match. You can use the * wildcard in the v attribute to match any value.

- The line element tells Osmarender to draw the selected features as lines.

- The style attribute of the line element contains CSS formatting instructions that will be included in the SVG file.

In addition to `line`, Osmarender supports the following rendering instructions:

- `area` draws any selected ways as a filled area, even if the way isn't explicitly closed. You can style the outline and fill of an area.

- `circle` draws a circle at the location of a node or at the center of a closed way. This is often used to note the location of a feature without having a special symbol for it.

- `symbol` draws a graphic symbol at the location of a node or at the center of a closed way. The symbol has to be an SVG file.

- `caption` draws text at the location of a node or at the center of a closed way.

- `pathText` is used to draw text along the path of a way, such as a road or waterway, rather than as a block caption.

- `wayMarker` draws symbols at a node, perpendicular to a way passing through that node. This is used to display features such as gates across roads.

Let's start a simple rules file that will produce similar results to the map we produced using Kosmos earlier in the chapter. The basic template for a rules file looks as follows:

```
<?xml version="1.0" encoding="UTF-8"?>
<?xml-stylesheet type="text/xsl" href="osmarender.xsl"?>

<rules xmlns:xlink="http://www.w3.org/1999/xlink"
       xmlns:svg="http://www.w3.org/2000/svg"
       data="data.osm"
       svgBaseProfile="full"
       scale="1"
       minimumMapWidth="0.5"
       minimumMapHeight="0.5"
       symbolScale="0.5"
       symbolsDir=".">

  <rule e="...">
    ...
  </rule>
  <defs>
    <style xmlns="http://www.w3.org/2000/svg"
           type="text/css">
      .map-background
      {
        fill: white;
        stroke: none;
      }
    </style>
  </defs>

</rules>
```

The first two lines are XML instructions and are crucial to the correct working of Osmarender, so don't leave them out.

The `rules` element contains all the instructions to Osmarender. It also has some options set in its attributes, and we've used a minimal set in our previous template. There are more options for Osmarender, such as the ability to add a border, grid, and scale to the map. We'll ignore these for now, but you can read the documentation for the options at `http://wiki.openstreetmap.org/wiki/Osmarender/Options`.

Osmarender draws the features in our map in the same order as the rules in the file, so put the rules for the features you want to appear behind others first, and the features you want on top last in the rules file.

The `defs` element in the rules is used to add further style information to the map, and the contents of this element are copied verbatim into the final SVG file for our map. In this case, the only style we're setting is for the background of the map, which we're setting to white. You can set the background to any color you like, but you may get odd results if you don't set one at all, depending on which SVG application you use.

Writing simple rules

Let's have a look at the rules that would produce similar results in Osmarender as we got from Kosmos earlier in the chapter. Open the file `osmarender-template.xml` in a text editor, and save it as `compton.xml` in the same folder as you've placed the Osmarender files.

To render the residential area of Compton, we need to select the way tagged as `landuse=residential` and render it as an area:

```
<rule e="way" k="landuse" v="residential">
  <area style="stroke: none; fill: #FFBBBB" />
</rule>
```

Here, you can see the styling we've given to the element. The `stroke` instruction of `none` tells the renderer not to draw any outline on the area. The `fill` instruction tells the renderer to use the same shade of pink that we used earlier.

We can already render our map based on just this rule. Save the file and open a command line window. To process our rules with Osmarender, issue the following command:

```
xmlstarlet tr osmarender.xsl compton.xml >compton.svg
```

If we now open `compton.svg` in a web browser, we should see something like the following image:

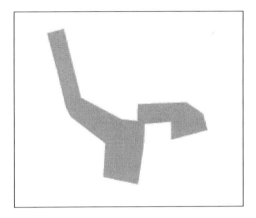

It's a good idea to run Osmarender and check the output after adding any rules to your file to make sure it's producing the effects you want.

Next, we render the unclassified and residential roads in the village. Unlike Kosmos, Osmarender won't draw any casings on roads for us, so to get the same effect, we have to draw a black line with a slightly narrower white one on top:

```
<rule e="way" k="highway" v="unclassified">
  <line style="stroke-width: 5px; stroke: black; fill: none" />
  <line style="stroke-width: 4px; stroke: white; fill: none" />
</rule>
<rule e="way" k="highway" v="residential">
  <line style="stroke-width: 4px; stroke: black; fill: none" />
  <line style="stroke-width: 3px; stroke: white; fill: none" />
</rule>
```

We've used two `line` instructions in each `rule` to draw a white road with black casing. We've also set the `fill` to `none` as a safety measure. Remember that Osmarender will fill any way if told to do so, not just closed ways, so explicitly making the `fill` of each way `none` will prevent this from happening. Remember that when drawing lines over the top of each other like this, features are drawn on top of each other in the order they appear in the rules file. If our two `line` instructions were the other way around, we'd just get a black line.

Next, let's render the trunk road and its link roads using the same style. This means we can use a single rule that matches both road types, which keeps our rules simpler:

```
<rule e="way" k="highway" v="trunk|trunk_link">
  <line style="stroke-width: 7px; stroke: black; fill: none" />
  <line style="stroke-width: 6px; stroke: #008000; fill: none" />
</rule>
```

Again, we've drawn our own casing. Finally, we draw the secondary roads:

```
<rule e="way" k="highway" v="secondary">
  <line style="stroke-width: 6px; stroke: black; fill: none" />
  <line style="stroke-width: 5px; stroke: #FDBF6F; fill:none" />
</rule>
```

We haven't yet drawn any road names, so let's add those:

```
<rule e="way" k="highway" v="secondary|unclassified|residential">
  <pathText k="name" startOffset="50%" dy="2.5px" style="font-family:
    Arial; font-weight: bold; font-size: 5px; text-anchor: middle"
    avoid-duplicates="true" />
</rule>
```

Here we've matched three road types in a single rule: secondary, unclassified, and residential. Note that we haven't used a wildcard as we don't' want to match the trunk road on our map. We've used a pathText instruction to draw the text along the line of the ways we've selected. This takes several parameters as attributes:

- The k attribute specifies the key of the tag whose value is to be used for the text. In this case it's the name tag.

- startOffset specifies how far along our way the text anchor should be placed. This is normally the start of the way, but we've said we want our text to be anchored halfway along each way. The alignment of the text against the anchor is specified in the style for the text.

- The dy attribute moves the text vertically from the line of the way. Normally, text is drawn with our way as its baseline, but we want the text roughly centered in the road.

- The style attribute again contains CSS formatting instructions for the text. Included in this is the text-anchor, which we've set to middle so that our text centers itself on the anchor point. Combined with the previous offset instruction, this places the text halfway along the road.

- The avoid-duplicated attribute tells Osmarender to avoid drawing a name twice wherever possible. In OpenStreetMap, a single road can comprise several ways, where each has the same name tag and each would get the name drawn on it

Finally, we need to mark the location of Compton Village Club. We'll use a star icon and text as before. However, we can't use the same image as before, as that was a PNG file and Osmarender can only use SVG graphics. We'll also add the name of the club in a caption:

```
<rule e="node" k="amenity" v="social_club">
  <symbol ref="star" position="center" />
  <caption k="name" dy="-1" style="font-family: Arial; font-weight:
    bold; font-size: 6px; text-anchor: middle;" />
</rule>
```

Here, we've selected the node representing the club and added two rendering instructions for it. The first draws the star:

- The `ref` attribute specifies the symbol file to use. This has to be an SVG file, and the `.svg` extension is automatically added to the filename. Osmarender looks for symbol files in the folder specified in its options using the `symbolsDir` attribute of the `rules` element, which we've set to the current directory. If your map uses symbols, you'll also need to add a `symbolScale` attribute to `rules` set this option, which we've set to `0.5`. If you omit either of these options, your symbols won't get included in your map.

- The `position` attribute specifies where relative to the node representing the feature the symbol should be placed. We've set this to `center`.

We've finished our rendering rules, so let's run Osmarender again and look at the results:

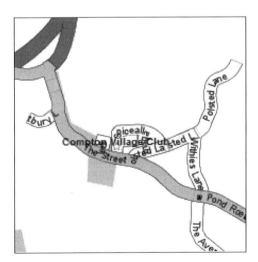

Although the features are all the right width and the text the right height, all the features are squashed together and the map doesn't look good. To space everything out, we use the `scale` option, which we originally set to 1. Altering the `scale` option changes the distance between features but not the effects of any rendering instructions, so the line widths and text sizes stay the same no matter what `scale` is set to.

Increase `scale` to 3 and rerun Osmarender. Refresh your view of the map file, and you should see something like the following screenshot:

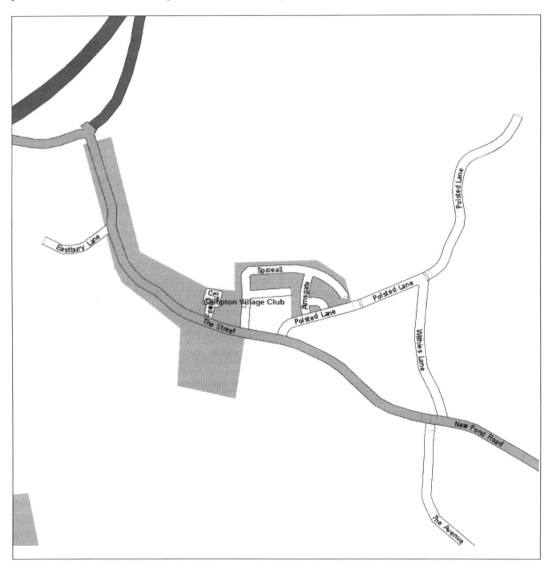

Everything is much more spaced out; the map itself is larger and easier to read. It's still not perfect, and some changes to the formatting would bring a further improvement in the appearance of the map. Even if we can't get Osmarender itself to produce the exactly results we're looking for, we can still edit the map by hand to put the finishing touches to it.

Using CSS classes for style

As we've already seen, you can include style information in a rules file that Osmarender will put in the finished SVG document. We can also use this to make our rules files simpler by putting all the formatting information for our map in **classes** and referencing those in our rendering rules. This has a couple of advantages:

- If there are features with different tags that we want to render in the same way, we don't have to write out all the formatting instructions each time

- All the formatting instructions are kept in a single part of the rules file, making it easier for a designer to edit the CSS without needing to understand the structure of a rules file

To make a rendering rule use a class, you add the formatting instructions to the style element in the defs section of your rules file. Choose a name for your class:

```
<defs>
  <style>
    .trunk-road
    {
      stroke-width: 7px;
      stroke: #008000;
      fill: none;
    }
    .trunk-road-casing
    {
      stroke-width: 13px;
      stroke: black;
      fill: none;
    }
  </style>
</defs>
```

Now, instead of adding a `style` attribute to the line for trunk roads, add a `class` attribute instead, with a value of our class name:

```
<rule e="way" k="highway" v="trunk|trunk_link">
  <line class="trunk-road-casing" />
  <line class="trunk-road" />
</rule>
```

Note that we can put the classes in the style section of the rules file in a different order to the rules that use them and still get the same results.

Nested rules

Osmarender allows you to nest rules to select features by more than one key. The best example of this is places of worship, where a different symbol is used depending on the religion, and perhaps even denomination, of the place of worship.

To select just a Christian place of worship so we can use a cross symbol to display it, we use two nested rules as shown in the following code:

```
<rule e="node|way" k="amenity" v="place_of_worship">
  <rule e="node|way" k="religion" v="christian">
    <symbol ref="church" />
  </rule>
</rule>
```

Here, in the preceding code, the outer rule selects any place of worship, whether mapped as a node or a closed way. The inner rule refines this by selecting only those features that are also tagged as Christian places of worship, and the rendering instruction inside this rule will only apply to those features. Note that you still have to include an e attribute for the inner rule, even though our outer rule has already selected the primitive types we're interested in. Without this attribute, our inner rule wouldn't select any features.

If we liked, we could nest rules for other religions inside the outer rule alongside our existing rule, as follows:

```
<rule e="node|way" k="amenity" v="place_of_worship">
  <rule e="node|way" k="religion" v="christian">
    <symbol ref="church" />
  </rule>
  <rule e="node|way" k="religion" v="jewish">
    <symbol ref="synagogue" />
  </rule>
  <rule e="node|way" k="religion" v="muslim">
    <symbol ref="mosque" />
  </rule>
</rule>
```

You can nest rules further than this if you need to, and this is used to a great effect in the rules for the Osmarender layer on openstreetmap.org.

Creating default rules using <else>

We've seen that we can nest rules to match more than one tag for a feature, but we can also create rules for features that don't match a particular tag using an `else` rule. What this allows you to do is use a different rendering for one special type of feature, and use a more general rendering for any other similar features.

For instance, if we wanted to use a special symbol to show the cafés that have WiFi available, but still show other cafés on our map, we could use the following rules:

```
<rule e="node|way" k="amenity" v="cafe">
  <rule e="node|way" k="wifi" v="yes">
    <symbol ref="cafe-with-wifi" />
  </rule>
  <else>
    <symbol ref="cafe" />
  </else>
</rule>
```

An `else` instruction only applies to features that do not match the rule immediately proceeding it, but do match any enclosing rule. This means you can't have a list of nested rules with a catch-all `else` rule at the end. So, in a set of rules like the following:

```
<rule e="node|way" k="amenity" v="place_of_worship">
  <rule e="node|way" k="religion" v="christian">
    <symbol ref="church" />
  </rule>
  <rule e="node|way" k="religion" v="jewish">
    <symbol ref="synagogue" />
  </rule>
  <rule e="node|way" k="religion" v="muslim">
    <symbol ref="mosque" />
  </rule>
  <else>
    <symbol ref="generic-pow" />
  </else>
</rule>
```

The `else` rule would apply to any features tagged as `amenity=place_of_worship`, but not tagged with `religion=muslim`, which isn't the effect you'd want to achieve. It would match all the Christian and Jewish places of worship as well. To render any place of worship we didn't have a specific symbol for using a generic symbol, we'd need to use rules like the following:

```
<rule e="node|way" k="amenity" v="place_of_worship">
  <rule e="node|way" k="religion" v="christian|jewish|muslim">
    <rule e="node|way" k="religion" v="christian">
      <symbol ref="church" />
    </rule>
    <rule e="node|way" k="religion" v="jewish">
      <symbol ref="synagogue" />
    </rule>
    <rule e="node|way" k="religion" v="muslim">
      <symbol ref="mosque" />
    </rule>
  </rule>
  <else>
    <symbol ref="generic-pow" />
  </else>
</rule>
```

Here, we've matched every value of `religion=*` for which we have a symbol in one rule and nested religion-specific rules inside that, and provided an `else` rule to match everything else.

More complex rules

We've just created a very simple map with Osmarender, but far more complex sets of rendering rules are possible. Some of the features used in the Osmarender layer on openstreetmap.org include:

- Pattern fills for areas
- Line cap styles
- Dashed and dotted lines
- Transparency

These are all features of CSS, and to get the most out of them you'll need a good knowledge of CSS. Although there is more documentation for Osmarender on the OpenStreetMap wiki, the best way of learning what's possible is to read more about CSS itself and how it applies to SVG, and by looking at the stylesheets used for the Osmarender layer on openstreetmap.org, all of which can be found at `http://svn.openstreetmap.org/applications/rendering/osmarender/stylesheets/`.

Summary

There are several ways of producing a single customized map from OpenStreetMap data, depending on which platform you want to use, the amount of customization you need, and your skill level:

- If you are happy with the standard map renderings, and don't need to add much extra information, just use the **exporter at openstreetmap.org**. This doesn't involve setting up any extra software, can be used from any computer, and requires little technical skill.

- **Kosmos** is a useful tool for Windows users. It allows you to render maps without extensive technical skills, and the rules are relatively easy to read on the wiki. You can use it to produce bitmap images of the size you want, in the style you want. However, it does have some bugs, only runs reliably on Windows, and isn't being developed any further.

- **Osmarender** is the most powerful and flexible renderer we've looked at here, but also the most difficult to operate and customize. It uses open technologies and tools and can produce very finely crafted results in the right hands, but needs a thorough knowledge of Cascading Style Sheets and Scalable Vector Graphics to get the very best out of it.

In the next chapter, we'll look at how you can use OpenStreetMap data in applications that don't understand the native OpenStreetMap XML format, which includes some viewing and rendering applications.

9

Getting Raw OpenStreetMap Data

We've looked at some small-scale uses of OpenStreetMap data covering a single village. This is fine up to a point, but you need larger amounts of data than this for any reasonably serious application. There are many ways of getting at the raw data in the OpenStreetMap database, depending on how much you want, how often you want it, and how much processing you're prepared to do yourself. The main methods are:

- Planet files
- The main OpenStreetMap API
- The Extended API (XAPI)

Each of these provides data in a different way, each with advantages and disadvantages. We'll discuss these in this chapter, and you should get an idea of which method will suit your application best.

As all of these methods provide data using Extensible Markup Language (**XML**), it will help you understand some of the examples given if you have a basic understanding of XML, but it is possible to retrieve, process, and use OpenStreetMap data without ever seeing a single line of XML, provided you use the right tools. This chapter only discusses ways of getting the data, and in the next chapter we'll look at some ways of post-processing this data into a form that's more suitable for using with applications.

Although you can use a normal web browser to get data from these sources, sometimes more data is returned than many browsers can cope with, and you'll probably want to do more with the data than just look at it. To retrieve data and store it in a file, you can use a command-line HTTP utility, of which wget is one of the most widely used. This program is included in most Linux distributions, and is available in virtually all Unix-like operating systems and Microsoft Windows. Installing wget on Linux is best done through the local package manager, and a Windows executable version is available at http://gnuwin32.sourceforge.net/packages/wget.htm. On Mac OS X, you can install wget from source, or via a third-party package manager such as MacPorts (http://www.macports.org/).

Planet files

A **Planet file** is a dump of the entire OpenStreetMap database to an XML file, containing the latest version of each feature in the world at the time it was produced. If you need OpenStreetMap data in significant quantities, downloading a Planet file is the best way of getting it, and for very large quantities of data, it is the only way of doing so.

We'll go through how you can obtain a planet file or other files based on it here. We'll cover the tools you can use to process a planet file in the next chapter.

The main Planet site

All Planet files are initially stored at http://planet.openstreetmap.org/. At present, one full planet file is produced every week, usually on Thursday morning GMT. Between full Planet files, a series of files containing the edits made to the database are produced, called **diffs**, short for differences, after the files used to change the source code of software. OpenStreetMap diffs use a customized file format, and you have to use tools specifically written for OpenStreetMap to process them.

As you can imagine, Planet files are massive. As of March 2010, the size for a compressed Planet file was 8.1 GB. Uncompressed, this would be around 160 GB, but fortunately most tools that process planet files can do so without decompressing the file. This means downloading a planet file using a standard web browser is impossible, so you'll need to use a utility such as wget to retrieve a planet file.

There's no fancy interface on the planet website, just a list of files and folders, so you need to know your way around. The planet files themselves are stored in the root folder, with a filename of the form `planet-yymmdd.osm.bz2`, where `yymmdd` is the date the file was produced. The latest planet file is also available as `planet-latest.osm.bz2`, so you can always get the latest full planet file using the URL, `http://planet.openstreetmap.org/planet-latest.osm.bz2`. Hence, to retrieve this using `wget`, you'd use the following command:

```
wget http://planet.openstreetmap.org/planet-latest.osm.bz2
```

Checking a planet file's integrity

For each planet file produced, an MD5 checksum is generated, which you can use to check the integrity of your download. This is highly recommended, as processing a planet file usually takes a long time, and won't necessarily fail straight away if your file is corrupted. The checksum file always has the same filename as the corresponding data file with `.md5` appended, so the checksum file for the latest planet file is at `http://planet.openstreetmap.org/planet-latest.osm.bz2.md5`. You can check the file for corruption by downloading the planet file and its checksum to the same folder, and running a checksum utility. On Linux, this is normally `md5sum`, and Microsoft provides a similar utility for Windows which you can download from `http://bit.ly/md5sumwin`.

To check the integrity of the file on Linux, use the following command:

```
md5sum -c planet-latest.osm.bz2.md5
```

This will check the file and give you an indication whether the file is OK.

On Windows, assuming you're using Microsoft's utility, use the following command:

```
fciv planet-latest.osm.bz2
```

This gives you the checksum for the file:

```
//
// File Checksum Integrity Verifier version 2.05.
//
68585722288ab8a4ad1ebe8a06c5f699 planet-latest.osm.bz2
```

Compare this against the one stored in the checksum file.

If you do find that your planet file download is corrupted, check on the OpenStreetMap wiki, the developer mailing list, or in the IRC channel to ensure the file on the server isn't corrupt before retrying.

Diff files

Besides the full planet files, other data is available for download from the planet website. This includes the diff files, dumps of changeset descriptions, and some experimental data file formats.

The most important of these are the diffs, which allow you to keep your local copy of the OpenStreetMap database up-to-date. As it's only practical to produce a planet file once a week, the changes made to the data in between are released as diff files, which contain the edits made in any one day, hour, or minute. By applying these changes to a planet file, you can keep it in sync with the true state of the database.

The diff files are in **osmChange** format (.osc) and are usually compressed using gzip to reduce their size, so a file will have a .osc.gz extension. The osmChange format is very similar to the standard OpenStreetMap XML format, but includes whether a feature is being added, changed, or deleted.

The hourly and minutely diffs contain every edit made during the period they cover, even if they cancel each other out. The daily diffs only contain the net results of the edits made during the period they cover at present, although this may change in the future. This doesn't really matter for users of the data, as either method will produce an up-to-date database.

The process of downloading and applying diffs can be completely automated using **Osmosis**—a tool for manipulating OpenStreetMap data in many formats. We'll cover how to do this in the next chapter.

There are some other files stored on the planet site, such as the experimental full history dumps. While you're welcome to download any of these files, they're mainly there to support development of new tools, and shouldn't be used in production applications yet.

Mirror sites

The central planet site can get overloaded, and as it's based in the UK, can be on the wrong side of a slow internet link for some people around the world. A series of mirror sites has been set up around the world to make access to the planet file quicker and more reliable. The full list of mirror sites is kept on the OpenStreetMap wiki at http://wiki.openstreetmap.org/wiki/Planet.osm#Mirrors, which includes both full mirrors of the planet site, and some that only carry the planet files but not other types.

Planet extracts

Not everyone wants or needs data for the entire planet, or for every type of feature. While it's possible to process the planet file using tools such as Osmosis to produce subsets of the data, it's much simpler to download a prepared extract of the planet. There are a number of sources for these, mostly from companies providing OpenStreetMap-related services, but some local OpenStreetMap organizations provide extracts for their own countries.

One source of extracts is German consultancy Geofabrik, which provides continent and country extracts of OpenStreetMap data at `http://download.geofabrik.de/osm/`. The data is provided in both native OpenStreetMap XML format and as ESRI Shapefiles. Not every country has its own file, and for some less-mapped regions, the only way of getting the data is through the continent extract.

CloudMade also provides extracts of OpenStreetMap data in native XML and other formats. These are available at `http://downloads.cloudmade.com/` and are again organized by continent. In addition to native format data, CloudMade provides ESRI Shapefiles, map files for Garmin GPS receivers, point of interest files for TomTom navigation devices, data files for Navit, as well as other subsets of the OpenStreetMap data.

Some local groups within OpenStreetMap produce their own extracts. At present, the Netherlands and Australia groups produce extracts covering the Benelux countries and Australia, respectively. For a comprehensive list of extract providers, refer to `http://wiki.openstreetmap.org/wiki/Planet.osm#Extracts`.

If you need an extract that's not already provided by one of these sources, you may be able to pay one of the commercial companies that provide OpenStreetMap services to produce an extract according to your exact requirements. Refer to `http://bit.ly/osmservices` for a list of companies offering such services.

Note that the license used by OpenStreetMap means that even if you have to pay to have a custom extract made, once you have the data, you're free to do whatever you like with it. Bear in mind that for areas that are being actively mapped, a single extract will go out of date very quickly.

OpenStreetMap's REST API

The main OpenStreetMap Application Programming Interface (API) is a web service that provides direct access to the OpenStreetMap database, and for virtually all purposes is the only way of accessing the database directly. All the editors we considered in Chapter 5 use this API to download and upload data.

The version of any feature you get from the main API is always the most up-to-date and is authoritative; that is to say, if there's a conflict between any OpenStreetMap data you have from another source and the API, you should always use the API data. The main API is also currently the only way of programmatically examining any feature's editing history and a particular user's actions, although development work is taking place to make this data more accessible.

Usage policy on the main API

The main OpenStreetMap API is provided to create and maintain the database, rather than to support any applications that use the data. The servers the API runs on are provided by the OpenStreetMap Foundation, supported only by donations. If you aren't planning to edit the database in any way, then you should avoid using the main API to get data that's available through other methods. Repeated use of the API to retrieve large amounts of data will result in you being blocked from using the OpenStreetMap servers, possibly without warning.

The API itself uses the **Representational State Transfer (REST)** style of web services. In a REST web service, each piece of data is given its own Uniform Resource Identifier (URI), better known as a web address, and all interaction with that data is through its URI. This contrasts with other styles of web services, such as the Simple Object Access Protocol (SOAP), where a single URI represents a function you want the service to perform. The main effect is that in REST web services you can normally rely on being able to link to a single piece of information. You can read more about REST programming at http://www.ibm.com/developerworks/webservices/library/ws-restful/.

It is possible to run your own copy of the OpenStreetMap API on a local server, and all the operations will be exactly the same bar the server name used. If you do run your own server, it won't contain the authoritative version of any feature, but you will get a much faster access to the data.

We're mostly going to cover the operations for reading data from the API in this chapter, although there are other operations that create, update, and delete data as well. In contrast to read operations, anything that alters the database in any way needs authenticating to a current openstreetmap.org account using a username and password, or via a web-based authentication system called OAuth.

The API also has a conflict detection system that prevents two mappers from editing the same feature at the same time, and this means it's not trivial to write data to the API by hand.

If you're thinking of writing an application that adds or changes data, you can read about the operations in the main API documentation at `http://wiki.openstreetmap.org/wiki/API_v0.6` for the current version. You should also contact the developer mailing list with your ideas to ensure you're not duplicating work already done, or planning to do something in a way that will overload the API servers.

Retrieving an individual feature

You can retrieve the data for any OpenStreetMap feature by accessing its URI using an HTTP request. The URI takes the form:

```
http://api.openstreetmap.org/api/<apiversion>/<type>/<id>
```

Here, `apiversion` is currently `0.6`, `type` is one of `node`, `way`, or `relation`, and `id` is the numerical ID of the element you want to retrieve. For instance, to get the node representing the village of Compton from the previous chapters using `wget`, you would use the following command:

```
wget -O compton-node.osm http://api.openstreetmap.org/api/0.6/
node/30361710
```

The `-O` option in the command line tells `wget` to save the returned document in a file named `compton-node.osm`, otherwise it would be saved in a file named `30361710`. This should produce an output like the following:

```
--2010-07-29 12:29:48--  http://api.openstreetmap.org/api/0.6/
node/30361710
Resolving api.openstreetmap.org... 128.40.168.98, 128.40.168.105
Connecting to api.openstreetmap.org|128.40.168.98|:80... connected.
HTTP request sent, awaiting response... 200 OK
Length: 335 [text/xml]
Saving to: `compton-node.osm'


100%[=====================================>] 335        --.-K/s   in 0s


2010-07-29 12:29:59 (25.2 MB/s) - `compton-node.osm' saved [335/335]
```

Similarly, to get the way representing The Street, the main road through Compton, you would use:

```
wget -O thestreet.osm http://api.openstreetmap.org/api/0.6/way/39429583
```

Finally, to get the relation we created earlier to group the separate parts of Spiceall, use:

```
wget -O spiceall.osm http://api.openstreetmap.org/api/0.6/relation/332067
```

For each of these calls, you get an XML file containing the details for the feature in question, but only that feature. For The Street, we get a file like the following:

```
<?xml version="1.0" encoding="UTF-8"?>
<osm version="0.6" generator="OpenStreetMap server">
  <way id="39429583" visible="true" timestamp="2009-11-20T10:44:55Z"
    version="4" changeset="3167314" user="Jonathan Bennett"
    uid="5352">
    <nd ref="27043724"/>
    <nd ref="81551586"/>
    .
    .
    .
    <nd ref="27075656"/>
    <tag k="highway" v="secondary"/>
    <tag k="maxspeed" v="30mph"/>
    <tag k="name" v="The Street"/>
    <tag k="ref" v="B3000"/>
  </way>
</osm>
```

This contains all the tags for The Street and a list of the nodes used, but no details of the nodes themselves. While each node could be retrieved using its own URL, this would be laborious and would create a disproportionate load on the servers. Instead, to make getting all the necessary data for a feature, there's a shortcut that lets you get all data associated with a feature. If you add /full to the end of a URI for a way or relation, all related nodes, and in the case of relations, ways will be returned. So, the following command returns every piece of data you need to draw Spiceall on a map:

```
wget -O spiceall-full.osm http://api.openstreetmap.org/api/0.6/
relation/332067/full
```

For nodes, there's no full option, as no other elements are needed for a feature described by a node. However, it is possible to get all the ways that use a particular node using a different call, by adding /ways to the end of a node URI, as follows:

```
wget -O spiceall-junction.osm http://www.openstreetmap.org/api/0.6/
node/27455427/ways
```

This returns the details of ways that use that node, in this case, the node at the junction of Spiceall and The Street. This isn't a full set of data, just the ways, so if you want to draw a map, you'll still need to fetch the full set of nodes for each way.

Similarly, there's a call that fetches all the relations that a feature is a member of, which uses a similar syntax:

```
wget -O spiceall-relations.osm http://api.openstreetmap.org/api/0.6/
way/4480421/relations
```

This will get the relations for one section of Spiceall, so in this case it will return our relation for the whole of Spiceall. As with the ways call, this only returns the relation itself, not any members.

Getting a feature's editing history

The OpenStreetMap database contains every version of every feature ever created, even those that have since been deleted. Normally, only the current version of any feature is returned by the API when you ask for it, and for drawing maps or any other normal use of the data, this is what you should use. However, if you're interested in when a particular tag was added to a feature or who added it, analyzing the history is the only accurate way to do it.

> Remember that a feature's editing history doesn't necessarily reflect the way the feature has changed in the real world. OpenStreetMap's crowdsourced approach to data means that any change is just as likely to be a mapper making the data more accurate as representing a change on the ground.

You can get the editing history for any individual feature by appending /history to its URI. Hence, for the editing history of The Street in Compton, you would use the following command:

```
wget -O TheStreet-history.osm http://api.openstreetmap.org/api/0.6/
way/39429583/history
```

This returns an OpenStreetMap XML file containing every version of this feature, each marked with a version number. The file will normally have the versions listed in ascending order, but you shouldn't assume this to be the case.

Each version of every feature has a `visible` attribute that in most cases will have a value of `true`. However, if the feature was deleted, the last version will have a value of `false`. Occasionally, a deleted feature may be restored.

You can also retrieve a specific version of a feature by appending its version number to the end of the URI, as shown in the following command:

```
wget -O TheStreet-v2.osm http://api.openstreetmap.org/pi/0.6/
way/39429583/2
```

This will get the data for the second version of the data for The Street. Note that again this only gets the data for the way, not its nodes.

There's also currently no simple way of getting the correct version of the nodes used by the version of the way you retrieve. You can get a list of every feature edited at the same time as your way, but that won't necessarily give you that information. Developers are working on possible solutions to this, but it may need to be implemented outside the main API for performance reasons.

Retrieving all features in an area

Getting the data for a single feature isn't normally that useful, so there's an API call that returns every feature for a given bounding box. This is known as the map call, and takes the following form:

```
http://api.openstreetmap.org/api/0.6/map?bbox=<left>,<bottom>,<right>
,<top>
```

Here `left`, `bottom`, `right`, and `top` are the edges of the bounding box given in decimal WGS84 degrees. Hence, to get every feature for the whole village of Compton, you'd use the following command:

```
wget -O compton.osm
http://api.openstreetmap.org/api/0.6/map?bbox=-0.64,51.21,-0.615,51.22
```

This call returns:

- All nodes within the bounding box
- All ways which have at least one node within the bounding box
- All remaining nodes for any of the ways included
- Any relations that have members within the bounding box, but not any members of that relation that wouldn't otherwise be included

This means certain features aren't returned that you might expect. These include the following:

- Any ways that cross the bounding box, but do not have a node within it
- The remaining members of any relation included in the bounding box

On the main OpenStreetMap servers, the maximum bounding box you can request using the map call is limited to an area of 0.25 square degrees, or a maximum of 50,000 nodes. If your request exceeds either of these limits, you'll get no results. If you need to retrieve data for a larger area than this, you will need to use the extended API, or XAPI.

What the map call also doesn't allow you to do is cross the antimeridian—that is, 180 degrees longitude—in a single query. In the unlikely event that you need a set of data from both sides of the antimeridian, you'll need to perform two separate queries and combine the two results.

The extended API (XAPI)

In addition to the normal OpenStreetmap API, there is an extended API (**XAPI**) that gives you the ability to query OpenStreetMap data for particular sets of data. For instance, you can request data that has specific tags, covers a specific area, or is used in a particular combination of primitive elements. It's the most versatile way of getting customized sets of data from OpenStreetMap, and if you don't have the resources to set up a local copy of the OpenStreetMap database, XAPI is your next best way of getting the data you need.

XAPI doesn't run on the main OpenStreetMap servers, but instead on separate machines, so that the main servers aren't overloaded. There are several instances of XAPI running at present, each with different update frequencies, availability, and capacity.

A complete list of current XAPI servers is kept on the OpenStreetmap wiki at `http://wiki.openstreetmap.org/wiki/XAPI`, including details of how often they're updated from the main database. The aim is to have XAPI serve data no more than 10 minutes old. However, as with some other OpenStreetmap-related services, XAPI is provided on a best-effort basis only, and you shouldn't assume that any or all of the XAPI servers will always be available or up-to-date.

XAPI queries are slow

Most non-trivial calls to XAPI will take several minutes to produce a result. This is caused by the technical requirements of XAPI itself and the hardware available to run it on. This means using a browser to perform XAPI queries is in most cases impractical, and you will need to use a command line utility, such as `wget`, if you're performing the queries by hand. If you are retrieving data from XAPI within an application you've written yourself, you need to ensure that any time-out for HTTP requests is set to a long enough period for the queries to complete.

Standard API calls

XAPI supports most of the basic API calls just covered. You can get the current version of any individual feature from XAPI in exactly the same way as the main API. Using the instance of XAPI hosted alongside the main servers, means we can get the data for The Street in Compton from the URI `http://xapi.openstreetmap.org/api/0.6/way/39429583`.

What you can't use are any of the options for a feature, such as the history or version calls; that information is only available through the main API. You automatically get the nodes for any ways and members for any relations your query returns, so the `/full` option is unnecessary.

Query by primitive

In addition to the standard API calls that retrieve specific features by numerical ID, XAPI can return every instance of each primitive type, normally used in conjunction with other filters. URIs for queries take a similar form to standard API calls, but omit the numerical ID from the path. In addition to `node`, `way`, and `relation`, you can also query XAPI for all matching features, irrespective of primitive using the `*` wildcard:

```
http://xapi.openstreetmap.org/api/0.6/*[...]
```

This will return any items in the database matching the filters you specify, including any other features they depend on, so for ways you get the required nodes, and for relations you get all members. This means you can normally use the results of an XAPI query directly in an application without needing to download any more data.

Map query

XAPI allows you to retrieve all data in a bounding box using the same call as the main API, but allows a far larger bounding box of up to 100 square degrees. Consider the following command:

```
wget http://xapi.openstreetmap.org/api/0.6/map?bbox=-25.09,62.94,-12.55,67.42
```

This will fetch all the data for Iceland, including any outlying islands. Retrieving this amount of data wouldn't be possible using the main API, and even using XAPI, it's not a quick operation. Running this command took just over an hour to retrieve the 88MB of data for Iceland when we tested it.

The map query doesn't support any further filtering of the data before it's returned. It's also possible to use a bounding box in addition to other filters in XAPI, but this has a slightly different syntax.

Filtering data by area

The previous map call is useful, but other than a larger area, it doesn't offer anything the main API doesn't. XAPI's power is in being able to combine a bounding box with other filters to get more specific sets of data.

The syntax of a bounding box filter is the same as that of a map query, except that the filter is surrounded by square braces. For example, to get the same data at the previous map query, you'd use the following command:

```
wget-O iceland.osm http://xapi.openstreetmap.org/api/0.6/*[bbox=-
25.09,62.94,-12.55,67.42]
```

This should return exactly the same set of data as the preceding map call, but also serve as a starting point for us to get more specific sets of data from XAPI. For instance, we can combine the same bounding box filter with a primitive query to retrieve just the ways in that area:

```
wget -O iceland-ways.osm http://xapi.openstreetmap.org/api/0.6/way[bbox=-
25.09,62.94,-12.55,67.42]
```

As previously noted, this will also return all the nodes needed to draw the ways. Any nodes used to map points of interest, any other unconnected nodes, and any relations will be omitted.

Filtering by tag

XAPI allows you to filter the data according to which tags features have. This is probably the most powerful part of XAPI, and the right combination of filters can get you exactly the right set of data for your map or application.

You can specify that any data returned must have a tag with a specific key, or a specific key and value. The syntax of a tag filter is key=value, surrounded by square braces, and value can take a wildcard value. XAPI used to support wildcards in the key of a tag, but support for this has been discontinued for performance reasons.

We can take our previous Iceland example further by querying for ways with specific tags. Any coastline should be made up of ways tagged with natural=coastline. We can get the data for Iceland's coastline by combining a tag filter with our bounding box filter, as follows:

```
wget -O iceland-coastline.osm http://xapi.openstreetmap.org/api/0.6/
way[bbox=-25.09,62.94,-12.55,67.42][natural=coastline]
```

After a short delay we get back the 14MB data of the coastline. This is a small enough amount of data for one of the desktop editors to handle, and following is what the coastline looks like in Merkaartor:

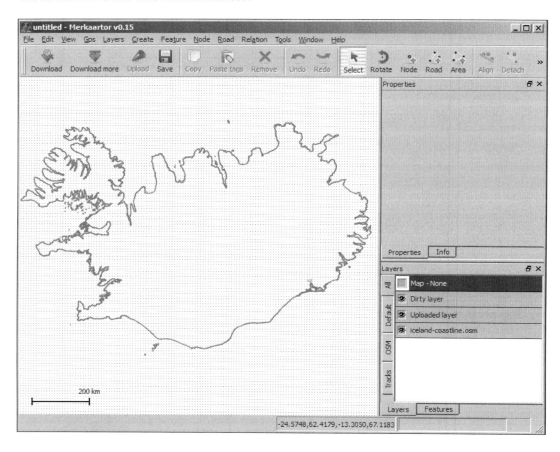

We could also get any road, footpath, and other route in Iceland by using a wildcard query, specifying any features tagged as `highway=*` within our bounding box.

```
wget -O iceland-roads.osm http://xapi.openstreetmap.org/api/0.6/
way[bbox=-25.09,62.94,-12.55,67.42][highway=*]
```

Tag filters can also match multiple values, separated with the bar character |. If we wanted just an overview of the major roads in Iceland, we could use a query like the following:

```
wget -O iceland-major-roads.osm http://xapi.openstreetmap.org/api/0.6/
way[bbox=-25.09,62.94,-12.55,67.42][highway=motorway\|trunk\|primary]
```

Note that when using wget under a Unix-like operating system, the bar characters need escaping using a backslash to prevent the shell from interpreting them as a Unix pipe command.

You can also match multiple keys using the bar character (|), although this is unlikely to produce the results you're looking for, except in a few specific circumstances. These tend to be where different groups of mappers have used different keys to map the same type of feature, although this doesn't happen very often, and is usually corrected when it does. Imagine that cinemas had been tagged using both amenity=cinema and leisure=cinema, and you needed all cinemas; you could use the filter [amenity|leisure=cinema] to find them.

Although you can match multiple values for a single key in a tag filter, you're only allowed one tag filter per query, which limits how specific a query can get. If you wanted to run a query to find every Synagogue in an area, ideally you'd use two tag filters to match both amenity=place_of_worship and religion=jewish, but this isn't possible. While you could just match the religion=jewish tag, this will also include any Jewish burial places that have been mapped, which may not be what you want. If you're getting data to render into a map, one workaround would be to use this filter anyway, then make sure your rendering rules ignore any features that don't have both tags.

Filtering by associated elements

A combination of element, bounding box, and tag filters should be able to give you most sets of data you're likely to want to render to a map or use in an application, but XAPI still has another type of filter that is particularly useful when trying to troubleshoot a particular set of data, the relationship the data has to other elements. The XAPI documentation calls these "child element" filters, but the description isn't entirely accurate.

The list of filters is as follows:

- /api/0.6/way[nd] selects ways that have at least one node
- /api/0.6/way[tag] selects ways that have at least one tag
- /api/0.6/way[not(nd)] only selects ways that do not have any nodes
- /api/0.6/way[not(tag)] only selects ways that do not have any tags
- /api/0.6/node[way] selects nodes that belong to at least one way
- /api/0.6/node[not(way)] selects nodes that do not belong to any way
- /api/0.6/node[not(tag)] selects nodes that do not have any tags

- `/api/0.6/relation[node]` selects relations that have at least one node member

- `/api/0.6/relation[way]` selects relations that have at least one way member

- `/api/0.6/relation[relation]` selects relations that have at least one relation member

- `/api/0.6/relation[not(node)]` selects relations that do not have any node members

- `/api/0.6/relation[not(way)]` selects relations that do not have any way members

- `/api/0.6/relation[not(relation)]` selects relations that do not have any relation members

Note the slight change in syntax between the `/way[nd]` and `/relation[node]` calls. Note also that if you're using a command-line client, such as `wget`, with these filters, you'll need to escape the parentheses with a backslash.

The most obvious use of these filters is to find incorrect data. Ways should always consist of two or more nodes, so the following call should never return anything:

```
http://xapi.openstreetmap.org/api/0.6/way[not(nd)]
```

If you run this query in conjunction with a bounding box, it will tell you of any data in that area that could cause rendering problems.

Some tags for nodes only make sense if the node is part of a way, such as `highway=mini_roundabout`, so any node that isn't part of a way should never have that tag. The following query will return any that do:

```
http://xapi.openstreetmap.org/api/0.6/node[not(way)][highway=mini_
roundabout]
```

Another example would be a turn restriction modeled using a relation. A turn restriction is tagged using `type=restriction`, so the following query will find any such relations without the node required:

```
http://xapi.openstreetmap.org/api/0.6/relation[not(node)][type=restri
ction]
```

Hopefully, this query returns no results, as there should be no such relations in the database.

Filter by user activity

One final way of filtering data in XAPI is by when and by whom it was last edited. Again, this is mostly of interest when you're troubleshooting data or managing vandalism in the OpenStreetMap database, rather than creating applications that use the data.

The user filters are [@user=<username>] or [@uid=<user id>] to find features that were last edited by a particular user, or [@changeset=<id>] to find features that were last edited in a particular changeset.

For instance, consider the following query:

```
wget -O jonathan-bennett.osm http://xapi.openstreetmap.org/api/0.6/*[@uid=5352]
```

This will return every feature in the OpenStreetMap database that was last edited by the author. This should be a range of features scattered around the UK.

Similarly, you can get any features last changed in changeset number 3265548, which makes a few changes in Compton using the following query:

```
wget -O changeset-3265548.osm http://xapi.openstreetmap.org/api/0.6/*[@changeset=3265548]
```

Note that while this will return a full set of data for the features in this changeset, including the nodes for any ways, they will be the latest version of those nodes, rather than the version that was current at the time of the changeset.

If you encounter edits to the OpenStreetMap that are clearly vandalism, you can use a combination of these queries to find other edits that the same mapper has made.

Summary

Given the number of options available, you should always be able to get OpenStreetMap data that's at most a week old. This contrasts with proprietary geodata providers where the data could be months, possibly years out-of-date, expensive, and with restrictive license conditions attached. Getting customized subsets of data would probably be an additional expense. With OpenStreetMap, it's easy and free.

Choose the method of getting data that suits your needs best:

- If you need country-sized sets of data, use the planet file or a pre-compiled extract file.
- If you only need specific sets of features or smaller areas, rather than a whole country, use XAPI.
- To get the most up-to-date version of a feature you know has changed recently, use the main API.

With OpenStreetMap, you get freedom to do things the way they suits you. You can download the whole dataset ready for processing, retrieve subsets of data for specific purposes, and even get individual features directly from the database. Even if you find an error in the data, you're still able to edit it yourself to correct the problem and share the fix with everyone else using the data.

In the next chapter, we'll look at one of the main tools used by OpenStreetMap itself and others using the data to transform it into more usable formats.

10
Manipulating OpenStreetMap Data using Osmosis

If you're going to use large amounts of OpenStreetMap data in a project, whether that's a large area or very detailed mapping of a smaller area, you'll probably need to change the data in some way before you can use it. Often this involves only using a subset of the data available, whether that's based on area, or the types of features included.

In this chapter, we'll look at a tool that is used heavily within OpenStreetMap to manipulate data from planet files or extracts, or databases containing OpenStreetMap data, called Osmosis.

What is Osmosis?

Osmosis is a command-line Java application that's often called the "Swiss Army Knife" of OpenStreetMap. It can perform a number of tasks to manipulate OpenStreetMap data in some way, and thanks to its plugin-based architecture, can be extended to perform new tasks. For many large-scale uses of OpenStreetMap data, Osmosis is the first tool you'll use before trying to render a map or use the data in some other way. For some tasks, it's the only way of getting the job done.

Some examples of what Osmosis can do include:

- Extracting data inside a bounding box or polygon
- Filtering the data based on primitive type and tags
- Splitting one large OpenStreetMap file into several smaller ones
- Importing data into an OpenStreetMap server
- Generating a list of changes between two OpenStreetMap files
- Applying diff files to a planet file or database to keep it up-to-date

We'll cover a simple example of each of these tasks.

Osmosis itself is open source software, released into the public domain. This means that not only can you use it without restriction, you can also incorporate the code into your own products without having to make the resulting software open source as well.

Setting up Osmosis

You can run Osmosis on any platform that has Java 1.6. To check which version of Java (if any) you have installed, use the following command:

```
java -version
```

If you don't have version 1.6 installed, you can download it for Windows and Linux from `http://www.java.com/`, and many Linux distributions also allow you to install Java via their package managers. On Mac OS X, you need to use the built-in software update to get the latest available version of Java. However, Java 1.6 is not available for PowerPC or early Intel-based Macs.

Our examples use version 0.35 of Osmosis, which you can download in either ZIP or GZIP format:

- `http://dev.openstreetmap.org/~bretth/osmosis-build/osmosis-bin-0.35.zip`
- `http://dev.openstreetmap.org/~bretth/osmosis-build/osmosis-bin-0.35.tgz`

These are binary-only builds of Osmosis and don't contain the source code. If you're interested in developing code for Osmosis or reusing its code in another application, you can download builds containing the source code, or check out the latest version of the source code from the repository. Refer to `http://wiki.openstreetmap.org/wiki/Osmosis#Downloading` for more details.

Once you've downloaded Osmosis, unpack the file into a suitable folder. Osmosis is a command-line utility, so you'll need a shell or command window open. Change to the directory you unpacked Osmosis into, then change into the `bin` subdirectory.

To check whether Osmosis is working, type the following command:

```
osmosis
```

You should see the following output:

```
02-May-2010 11:42:38 org.openstreetmap.osmosis.core.Osmosis run
INFO: Osmosis Version 0.35
```

```
02-May-2010 11:42:38 org.openstreetmap.osmosis.core.Osmosis run
INFO: Preparing pipeline.
02-May-2010 11:42:38 org.openstreetmap.osmosis.core.Osmosis run
INFO: Launching pipeline execution.
02-May-2010 11:42:38 org.openstreetmap.osmosis.core.Osmosis run
INFO: Pipeline executing, waiting for completion.
02-May-2010 11:42:38 org.openstreetmap.osmosis.core.Osmosis run
INFO: Pipeline complete.
02-May-2010 11:42:38 org.openstreetmap.osmosis.core.Osmosis run
INFO: Total execution time: 360 milliseconds.
```

On OS X and some Linux and Unix systems, you may need to use `./osmosis` on the command line.

How Osmosis processes data

Osmosis takes a list of command-line arguments that tell it what **tasks** to perform. Each task in turn will usually have its own arguments, and a complete command line for Osmosis can get very long. As a result of this, each task has both a long and short form, with the latter being easier to type, but harder to read if you can't remember each command. We'll use the long form in all our examples, so that it's obvious what we're doing. Once you're more familiar with Osmosis, you can find all the short versions of each task documented at `http://wiki.openstreetmap.org/wiki/Osmosis/Detailed_Usage`.

Command-line arguments are case-sensitive

All the options and parameters used by Osmosis on the command line are case-sensitive, even on Windows, so make sure that you include capital letters where indicated.

Always use the scripts located in the `bin` subfolder to start Osmosis. Trying to use the Java files directly won't produce the results you're looking for in most cases.

Osmosis uses a **pipeline** to process data. The pipeline has a number of stages, consisting of one or more input tasks, one or more optional processing tasks, and one or more output tasks. For some complex jobs, Osmosis may have more than one pipeline running simultaneously. In addition to carrying OpenStreetMap data, a pipeline can carry changes to the data.

To import data from a file into a database, for example, only an input task to read the XML and one output task to write to the database are used. To cut out a bounding box from a file needs an input task, the bounding box task, and an output task. Osmosis works by feeding the results of one task into another.

In most cases, the tasks have an inherent logical order, but for more complex combinations of tasks, we might want to change the way the tasks are linked together. Osmosis allows you to do that, and we'll cover how to change the pipeline order later in the chapter.

Processing planet files in Osmosis

The planet files produced each week are compressed using an advanced version of the `bzip2` utility, called `pbzip2`, which can process more than one data stream at a time, reducing the amount of time needed to produce each planet file. This means that Osmosis' built-in bzip decompression routines can't handle the files. We'll show you how to work around this issue, but remember, you can't process planet files using Osmosis on its own.

Bear in mind that any task in Osmosis that deals with any significant amount of data, and particularly entire planet files, will take minutes or hours to complete, even on very powerful computers. Osmosis also gives you very little feedback while it's running to show you what's happening, so don't panic if after issuing an Osmosis command, it looks like it's frozen. Be patient, and you should get your results, or an error message on screen if it can't be completed.

Cutting out a bounding box

One of the simplest tasks is to cut out a bounding box from a planet (or other) file to get data for a specific region. Let's use Iceland as our example, as we did with XAPI in the previous chapter. To extract the data of Iceland from a planet file, use the following command:

```
bzcat planet-100428.osm.bz2 | osmosis --read-xml file=/dev/stdin --
bounding-box left="-25.09" right="-12.55" top="67.42" bottom="62.94" --
write-xml file="iceland.osm.bz2"
```

The command uses three tasks to process the data. They are as follows:

- Osmosis reads the data in from a planet file from April 28, 2010 using the `--read-xml` task.

- It then extracts a bounding box using the `--bounding-box` task. As with other OpenStreetMap tools, the coordinates for the bounding box are supplied in WGS84 degrees.

- Finally, it writes the results to a file named `iceland.osm.bz2`, using the `--write-xml` task, compressing the output.

Normally, Osmosis works directly with bzip2-compressed files, and is able to detect automatically the format of any data files based on the file's extension. You can specify the compression used in any file if it has a non-standard extension, using the `compressionMethod` parameter, which can take `none`, `gzip`, or `bzip2` as values.

In the case of the planet file, you need to use the separate `bzcat` utility, as the planet files use an advanced form of compression that Osmosis's built-in decompression code can't handle. This utility is available on most Linux distributions, and you can download a Windows version from `http://gnuwin32.sourceforge.net/packages/bzip2.htm`. You can manually compress and decompress any files you deal with, and in the case of the planet file itself, you'll need a lot of disk space to work with the uncompressed data.

Getting data from outside the bounding box

By default, the bounding box task will only return nodes inside the specified bounding box. If there are ways partially outside the bounding box, any nodes that those ways use from outside the bounding box won't be included. This can cause problems with some software that uses OpenStreetMap data. To get all nodes used by ways that intersect the bounding box, use the bounding box task's `completeWays` option, as follows:

```
bzcat planet-100428.osm.bz2 | osmosis --read-xml file=/dev/stdin --
bounding-box left="-25.09" right="-12.55" top="67.42" bottom="62.94"
completeWays="yes" --write-xml file="iceland-complete.osm.bz2"
```

This will return a complete set of data that shouldn't cause any problems in any software, but will contain data from outside the specified bounding box.

There's also a `completeRelations` option that will return all members of any relation that has a member inside the bounding box, but you need to use this with more caution. If any large relations cross your bounding box, such as long-distance routes, you could end up with data far outside your original bounding box.

Cutting out data with a bounding polygon

Iceland is convenient to cut out using a simple bounding box, as it's surrounded by sea and has enough space around it that the box doesn't touch any other land. For most countries, you'll need to cut out a more complex shape from a planet file, and that's where the bounding polygon task comes in.

The bounding polygon task works in a similar way as the bounding box task, but instead uses a set of arbitrary polygons to define the area for which you want data. This is particularly useful where you're trying to cut out a country that shares land borders with others, or where a coastal boundary doesn't form a straight line. The bounding polygon task is also much slower than a bounding box, as a more complicated algorithm is needed to decide whether a point lies inside the polygon. This may not be a problem, but it's worth bearing in mind.

Using polygon files

The bounding polygon task needs a file describing the polygon you want to use in a special format. The file format is described at `http://wiki.openstreetmap.org/wiki/Osmosis/Polygon_Filter_File_Format`, but essentially consists of a list of coordinates of each corner of the polygon. It's also possible to define areas to subtract from the polygon, so that you can exclude enclaves from a country, for example.

Fortunately, we don't have to build the polygon files from scratch, as files for most countries around the world are already available from various sources. We're going to download a polygon file of the UK from Cloudmade. Download `http://downloads.cloudmade.com/europe/united_kingdom/united_kingdom.poly` to the `bin` subfolder of your Osmosis installation.

Now, to extract data for the UK from a planet file, we use the following command:

```
bzcat planet-100428.osm.bz2 | osmosis --read-xml file=/dev/stdin --
bounding-polygon file="united_kingdom.poly" --write-xml file="united_
kingdom.osm.bz2"
```

This will give us data for the whole of the United Kingdom, including Northern Ireland.

We'll use this file in the rest of our examples for this chapter, so you can either try extracting your own, or downloading a prepared extract from Cloudmade at `http://downloads.cloudmade.com/europe/united_kingdom/united_kingdom.osm.bz2`.

As with the bounding box task, the bounding polygon task can take `completeWays` and `completeRelations` options.

Creating your own polygon files

If you need to create your own polygon file to extract an area not already covered, you can code one by hand in the format documented at the previous link, or an easier way is to draw the polygons you want using JOSM, and convert them to the required format. To do this, you'll need a computer with Perl installed. We're going to create a polygon for the Isle of Wight in the UK. This was one of the first places to be mapped comprehensively in OpenStreetMap, and it can't be cut out using a simple bounding box without including some of the mainland.

First, download the script that converts OSM data to polygon files from http://svn. openstreetmap.org/applications/utils/osm-extract/polygons/osm2poly.pl.

Run JOSM as normal, and download the data for the area around which you want to create a polygon to use as a guide. If you want to extract a very large area, this could mean downloading the data in several chunks. The Isle of Wight will need two separate downloads.

Once you have the data of the area you're interested in, choose **New** from the **File** menu in JOSM. This will create a new data layer, and the existing data should now be grayed out. Draw a closed way around the area you want to cut out, making sure you end on the node on which you started. JOSM should stop drawing when you do. You can draw more than one way, as long as all the ways you draw are closed.

For the Isle of Wight, you should end up with something like the following image:

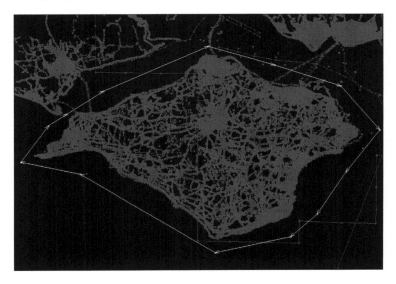

Once you have your area as one or more polygons, right-click on the top data layer in JOSM's Layers panel and choose **Save** from the pop-up menu. Choose a descriptive name for the file; in this case, `isle-of-wight.osm` will do.

Next, we need to convert the OpenStreetMap file to a polygon file using the utility we downloaded. With the `.osm` file in the same directory as the script, use the following command to convert the file:

```
perl osm2poly.pl isle-of-wight.osm >isle-of-wight.poly
```

The script prints its results directly, so you need to direct it to the filename of your choice. The script doesn't print anything on screen while it's running, and once it has completed the conversion, you should have a file with contents similar to the following:

```
polygon
1
    -1.298978E+00      5.077497E+01
    -1.095930E+00      5.073521E+01
    -1.082186E+00      5.072270E+01
    -1.041225E+00      5.068537E+01
    -1.166828E+00      5.059231E+01
    -1.302833E+00      5.056872E+01
    -1.499613E+00      5.065986E+01
    -1.579896E+00      5.065564E+01
    -1.598057E+00      5.066019E+01
    -1.533299E+00      5.070682E+01
    -1.480437E+00      5.073020E+01
    -1.298978E+00      5.077497E+01
END
END
```

Here, you can see the coordinates of each point in our bounding polygon expressed in decimal degrees.

You can also use data from OpenStreetMap to define a polygon, particularly political or administrative boundaries. We could use the Isle of Wight's coastline to define our polygon, although this is far more complex than the polygon we just created and would require us to create a single way from the several used to map the coastline.

Using tag filters to produce tailored datasets

You can extract particular types of features from OpenStreetMap data to give you a dataset that only includes the features you're interested in, and not others. You can filter the data based purely on a combination of primitive type and tags, and you can combine multiple filters to produce a set of data to fit your exact requirements.

The tag filter task works on a single primitive type per task, and can include or exclude those primitives from the pipeline based on the tags used. It leaves other types of primitive untouched.

To extract all nodes from our United Kingdom extract tagged with shop=*, we'd use the following command:

```
osmosis --read-xml file="united_kingdom.osm.bz2" --tag-filter accept-
nodes shop=* --tag-filter reject-ways --tag-filter  reject-relations --
write-xml file="uk-shops.osm"
```

Already, you can see multiple tasks in use in Osmosis. We're using three separate tag filter tasks:

- The first filters the nodes in the pipeline to include only those with a shop=* tag
- The second filters out all ways
- The last filters out all relations

This is also a good example of Osmosis's pipeline in action. The data is read from our extract into the pipeline, then passes through each of our filter tasks, and finally is written out to an XML file. We don't have to tell Osmosis how to direct data between the separate tasks, as they form a single pipeline. We could execute the filter tasks in any order and get the same results, but this isn't true for every combination of filters.

Note that because we didn't add a `.bz2` extension to the output file, Osmosis won't compress the data. We can open this file in JOSM, as follows to see where all the shops mapped in the UK are:

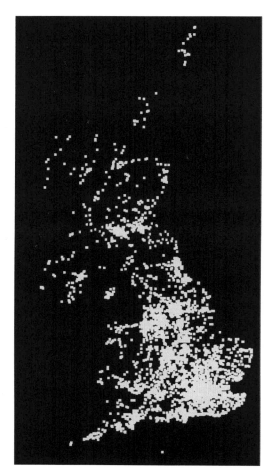

You can use the same type of filter more than once in a pipeline using multiple tasks, so to get all roads in the UK where you are allowed to cycle, you would use a series of filters like the following:

```
osmosis --read-xml file="united_kingdom.osm.bz2"
--tag-filter accept-ways highway=*
--tag-filter reject-ways bicycle=no
--tag-filter reject-ways highway=motorway,motorway_link,footway
--used-node
--tag-filter reject-relations
--write-xml file="uk-cycle-roads.osm.bz2"
```

Here we have a complex series of filters, and following what they do:

- The first filter (`--tag-filter accept-ways highway=*`) only allows through ways tagged with `highway=*` to remove anything that's not a road or path.

- The second filter (`--tag-filter reject-ways bicycle=no`) filters out ways tagged with `bicycle=no` to remove any roads where bicycles are explicitly not allowed.

- The third filter (`--tag-filter reject-ways highway=motorway,motorway_link,footway`) filters out roads tagged as motorways, motorway access roads, or footpaths. Here we see a rule matching multiple values for a tag by including a comma-separated list of values.

- The fourth filter (`--used-node`) isn't a tag filter, but only allows through nodes that are used by any ways that have made it through any preceding filters.

- The last filter removes all relations from the pipeline.

The result is a set of data representing any road or path in the UK on which it's legal to cycle. You could then feed this information into a routing engine, or use it to render a map specifically for cyclists.

Simplifying filters

There are also a set of combined tag and primitive type filters that have a shorter syntax, but are less flexible, and only work on nodes and ways, not relations. They also remove any other type of primitive from the pipeline, unlike the longer tag filter task. The filters available are:

- `node-key`
- `node-key-value`
- `way-key`
- `way-key-value`

Each of these filters only allows through primitives of the matching type, with the specified key(s) or key-value pair(s). To extract every node tagged as `shop=bicycle` in our UK extract, we use the following command:

```
osmosis --read-xml file="united_kingdom.osm.bz2" --node-key-value
keyValueList="shop.bicycle" --write-xml file="uk-bike-shops.osm"
```

The preceding command uses a filter that matches only nodes with tags matching the list in the `keyValueList` parameter. In our case, we've set `keyValueList` to a single pair of `shop=bicycle`, but note that a period is used to separate the key and value, rather than the equals sign.

This is equivalent to a set of tag filters, as shown in the following command:

```
osmosis --read-xml file="united_kingdom.osm.bz2" -tag-filter accept-nodes
keyValueList="shop.bicycle" --tag-filter reject-ways --tag-filter reject-
relations --write-xml file="uk-bike-shops.osm"
```

To specify more than one tag, we separate the key-value pairs with commas, so we can match both bicycle shops and bicycle parking with a single filter, as follows:

```
osmosis --read-xml file="united_kingdom.osm.bz2" --node-key-value
keyValueList="shop.bicycle,amenity.bicycle_parking" --write-xml file="uk-
bike-facilities.osm"
```

This gives us a set of places in the UK that might be of interest to cyclists while they're out and about on their bikes. We could extend this to include convenience stores, cafés, or other places by adding their tags to the list.

The right combination of tag filters can produce any imaginable subset of OpenStreetMap's data, even if it can't be done in a single operation. Next, we'll look at how you can produce parts of your desired dataset separately, then combine them into a single file.

Splitting and merging data streams

In our previous examples, we've produced two sets of data of interest to cyclists in the UK, namely, the roads on which they're allowed to cycle, and a list of places at which they might need to stop along the way. However, they're in two separate files, and the filters we've used to produce the sets of data are mutually exclusive.

Fortunately, we can merge the two sets into one quite simply, using Osmosis's **merge** task:

```
osmosis --read-xml file="uk-cycle-roads.osm.bz2" --read-xml file="uk-
bike-facilities.osm" --merge --write-xml file="uk-cycle-map.osm.bz2"
```

Here, we can see the two OpenStreetMap files being read in, each creating its own pipeline. These pipelines are then merged, and the results are written to another OpenStreetMap XML file. We can now use the result to render a map, or use it in another cycling-focused application.

This looks simple, but we've cheated slightly by using the data we've already generated using Osmosis. For the merge to work, the data in the source OpenStreetMap files has to be sorted, first by primitive type, then by ID. This should be the case with the data produced by Osmosis, but if you're unsure, or you're using data that's been generated by a different application, you can get Osmosis to sort the data prior to any merge operation.

To sort the data in a file, use the following **sort** task:

```
osmosis --read-xml file="unsorted.osm" --sort --write-xml file="sorted.osm"
```

This will produce a file in the correct order for Osmosis to merge with another set of data. While sort can take a `type` parameter, the only currently supported value is a sort by type, then ID.

The merge task only has two input pipelines, so if you need to merge more than two sets of data, you'll need to use multiple merge tasks, as shown in the following command:

```
osmosis --read-xml file="file1.osm" --read-xml file="file2.osm" --read-xml file="file3.osm" --merge --merge --write-xml file="combined.osm"
```

Creating multiple pipelines with the tee task

The merge task has a counterpart—the **tee** task. This produces multiple pipelines from the same input, so you can apply mutually exclusive sets of filters to a single source of data without having to run Osmosis several times.

Imagine we had our UK extract, and wanted to produce further extracts for each of the constituent countries of the UK, namely, England, Scotland, Wales, and Northern Ireland. We could run Osmosis with a bounding polygon task for each of these areas, but it's more efficient to do it in one operation. We can do it using the following command:

```
osmosis --read-xml file="united_kingdom.osm.bz2" \
    --tee outputCount="4" \
    --bounding-polygon file="england.poly" completeWays="yes" \
    --write-xml file="england.osm.bz2" \
    --bounding-polygon file="scotland.poly" completeWays="yes" \
    --write-xml file="scotland.osm.bz2" \
    --bounding-polygon file="wales.poly" completeWays="yes" \
    --write-xml file="wales.osm.bz2" \
    --bounding-polygon file="northern_ireland.poly" completeWays="yes" \
    --write-xml file="northern_ireland.osm.bz2"
```

In the preceding command, the `outputCount` parameter tells the tee task how many pipelines to produce. Unlike the merge task, a tee can produce many pipelines in a single operation. We then process each pipeline in its entirety before moving onto the next, so we have a bounding polygon task, then a write XML task together for each country. As England has land borders with Scotland and Wales, and Northern Ireland has a land border with the Republic of Ireland, we use the `completeWays` option to ensure that we include whole roads or paths that cross those borders.

The next logical step in producing our cycle map data would be to use a tee to split our UK extract, process the two pipelines to produce the highways and facilities separately, and recombine them in a merge task to produce our final file. Unfortunately, we can't do this, as, going by the way Osmosis works, trying to merge the pipelines of a tee will probably result in a deadlock at some point, as one portion of the pipeline waits for input that will never come.

To work around this issue, we can produce the two files in one operation, and merge them in another run of Osmosis. For the first operation, use the following command:

```
osmosis --read-xml file="united_kingdom.osm.bz2" \
    --tee \
    --tag-filter accept-ways highway=* \
    --tag-filter reject-ways bicycle=no \
    --tag-filter reject-ways  highway=motorway,motorway_link,footway \
    --used-node \
    --tag-filter reject-relations \
    --write-xml file="uk-cycle-roads.osm.bz2" \
    --node-key-value keyValueList="shop.bicycle,amenity.bicycle_parking" \
    --write-xml file="uk-bike-facilities.osm"
```

This is pretty long-winded, but does what we want. We can shorten the command line by using the shorter form of each command, which would look as follows:

```
osmosis --rx united_kingdom.osm.bz2
    --t
    --tf accept-ways highway=*
    --tf reject-ways bicycle=no
    --tf reject-ways highway=motorway,motorway_link,footway
    --un
    --tf reject-relations
    --wx uk-cycle-roads.osm.bz2
    --nkv keyValueList="shop.bicycle,amenity.bicycle_parking"
    --wx uk-bike-facilities.osm
```

You can see the short form of each task command, and that for file operations, we can use just the filename as a parameter. This makes the whole command line much shorter, but less readable if you don't know the options very well.

Once we have our two files, we merge them to produce our final UK cycling map dataset, as follows:

```
osmosis --read-xml file="uk-cycle-roads.osm.bz2" --read-xml file="uk-bike-facilities.osm" --merge --write-xml file="uk-cycle-map.osm.bz2"
```

We can then use `uk-cycle-map.osm.bz2` as input to a renderer or some other application.

Automatically updating local data from diffs

Planet files take a long time to download and process, so if you can avoid doing it every week, you'll save yourself a lot of time and bandwidth. Fortunately, Osmosis can help you by automating much of the work involved using its replication tasks.

Given a suitably configured system, Osmosis will download and apply all the changes to the data since a planet file was created, creating a copy that's almost as up-to-date as the main OpenStreetMap database. Run the same commands regularly, and, in theory, you will never need to download another planet file.

In this section, we'll see how to:

- Prepare your system for replication
- Run the initial update
- Schedule further updates to your data

Preparing your system

Before you can update a planet file, you need to have done some preparation work. Osmosis stores information about the current state of your data in a few text files. We need a working directory in which to store the files, so let's create it. Open the command line window and change to a folder, using the following command:

```
mkdir replication
```

Now we need to create the files Osmosis needs for replication, using the `--read-replication-interval-init` task:

```
osmosis --read-replication-interval-init workingDirectory="replication"
```

This creates two files in the `replication` directory, named `configuration.txt` and `download.lock`.

You don't need to alter these files to download data from OpenStreetMap, but if you have a set of diff files stored in a different place, or want to use a mirror site, you need to edit `configuration.txt` to reflect this. If you have more than one copy of the data, you'll need a separate working directory for each one, as you need to track the state of each copy separately.

We also need to tell Osmosis from what point to start downloading changes. This isn't simple with the replication diffs currently used, but you only need to do it once:

1. Download the latest planet file from `http://planet.openstreetmap.org/` if you don't already have it.

2. Note the timestamp of the planet file. The filename is of the form `planet-yymmdd.osm.bz2`, and in our examples we'll use a planet file from April 10, 2010.

3. Open `http://planet.openstreetmap.org/minute-replicate/` in a web browser.

4. You'll see a list of subfolders with a date next to each one, as shown in the following screenshot. Find the folder with a date similar to your planet, or with a date that is closest to your planet, but after it. In our case, this would be **300**.

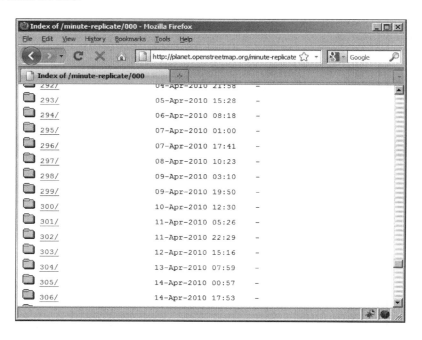

5. In that folder, find a file named **nnn.state.txt** with a timestamp just earlier than your planet file. Don't worry about choosing a file from earlier than your planet file, as any changes prior to your planet won't have any net effect on the data. We'll use `248.state.txt` in our example, as follows as this was the last diff produced on the day before our planet file:

247.state.txt	09-Apr-2010 23:58	184
248.osc.gz	09-Apr-2010 23:59	24K
248.state.txt	09-Apr-2010 23:59	164
249.osc.gz	10-Apr-2010 00:00	12K
249.state.txt	10-Apr-2010 00:00	194
250.osc.gz	10-Apr-2010 00:01	13K
250.state.txt	10-Apr-2010 00:01	184

6. Download this file and place it in your working folder. Rename the file as `state.txt`.

Your working folder should now have files named `configuration.txt`, `download.lock`, and `state.txt` in it. If it does, we're now ready to run the first update against our planet file.

Running the initial update

Once we have Osmosis configured, we can tell it to start downloading the diff files. We do this using the read replication interval task. We also then want to do something with the changes, which is usually to apply them to an existing set of data somewhere. In our case, that's an existing planet file. This will need three separate tasks:

- A read replication interval task to download and collate diff files from the OpenStreetMap servers

- A simplify change task to merge multiple changes to a single object into one operation

- An apply changes task to patch our planet file using the resulting changes

We could perform each of these tasks in a separate Osmosis operation using an intermediate XML file containing the changes, but we'll use Osmosis's pipelining to do everything in one command, as follows:

```
bzcat planet-100428.osm.bz2 | osmosis --read-xml file="/dev/stdin"
outPipe.0="planet" --read-replication-interval  workingDirectory="repli
cation" --simplify-change --apply-change inPipe.0="planet" --write-xml
file="/dev/stdout" | bzip2 >updated-planet.osm.bz2
```

This operation will take a very long time to complete, as is usual with operations on an entire planet file. You need to note how long this process takes, as it will tell you how often you'll be able to run a planet update. We're using the native bzip utilities rather than Osmosis's own compression routines to speed up the process, and if you have a machine with more than one processor core, this will significantly reduce the amount of time needed to patch the planet. If you have a large amount of disk space needed to store an uncompressed planet file—of the order of several hundred gigabytes—you can save even more time, as it's the file compression that takes most of the time.

Once the process is complete, you will have a file, `updated-planet.osm.bz2`, which will reflect the state of the main OpenStreetMap database at a time not much before your update operation started. In addition to updating the data, Osmosis will also record the current state of your data in `state.txt`, so that next time you run an update, it only downloads and applies the diffs produced since you last ran an update.

Keeping the data up-to-date automatically

Once you've applied the initial updates, you can keep your copy of the data up-to-date using a command similar to the one we used for the initial updates. You can even schedule the updates using your operating system's scheduler, so that your data is constantly kept up-to-date.

The command we use to keep our data updated is virtually identical to our initial update, but we use our updated planet as the input and a temporary file as the output:

```
bzcat planet-updated.osm.bz2 | osmosis --read-xml file="/dev/stdin"
outPipe.0="planet" --read-replication-interval  workingDirectory="repli
cation" --simplify-change --apply-change inPipe.0="planet" --write-xml
file="/dev/stdout" | bzip2 >planet-temp.osm.bz2
```

Once the update completes successfully, you can delete your old planet file and rename your temporary file.

On Unix and Linux, use the following commands:

```
rm planet-updated.osm.bz2
mv planet-temp.osm.bz2 planet-updated.osm.bz2
```

Use the following commands on Windows:

```
del planet-updated.osm.bz2
rename planet-temp.osm.bz2 planet-updated.osm.bz2
```

On Linux and other Unix-like operating systems, you can schedule the update using cron, and on Windows using the Windows Scheduler service. Remember that as updating an entire planet can take a very long time, you may only be able to perform the update once a day.

Reading the OpenStreetMap API from Osmosis

So far we've only dealt with data in OpenStreetMap XML files, but Osmosis can handle other sources of data, including databases and the OpenStreetMap API itself. Note that this is **not** the same as reading and writing to a database in API format, which we'll cover later.

Osmosis only supports the map call of the API to download data. We can get the data of Compton—the village we mapped in earlier chapters—and save it to a file using the read API task, as follows:

```
osmosis --read-api left="-0.64" bottom="51.21" right="-0.615" top="51.22"
--write-xml file="compton.osm"
```

We supply the bounding box we want to download as four parameters to the read API task.

However, remember that the main API only allows you to download a limited area, so if you need a large area, this method won't work. Fortunately, you can also get Osmosis to retrieve data from XAPI by supplying its URL as another parameter. If we wanted to retrieve data of Iceland from XAPI within Osmosis, we could do it using the following command:

```
osmosis --read-api left="-25.09" right="-12.55" top="67.42"
bottom="62.94" url="http://xapi.openstreetmap.org/api/0.6" --write-xml
file="iceland.osm.bz2"
```

Here, we add the URL of XAPI including the version, but without the trailing slash, as a parameter to the read API task.

Obviously, you don't need to write the data directly to a file. You can process it using tag filters, further bounding boxes or polygons, and use any other Osmosis tasks available.

Osmosis also supports uploading changes to the OpenStreetMap API, but there are few circumstances where you'd need to do this; reversing damage to the data (either accidental or intentional) is usually the only time you'd work this way. Any changes you need to make to OpenStreetMap's data should normally be done using one of the editors available.

Using Osmosis with a database

So far, we've only used Osmosis to process OpenStreetMap data in local XML files. It can also read and write data to a database, which is more complex to set up, but faster for many operations. Setting up a database for the data can be complicated, but you only need to do it once.

Osmosis supports three types of databases:

- PostGIS with a simple schema
- The OpenStreetMap API schema on PostgreSQL
- A legacy API schema on MySQL

We'll cover the first of these in this chapter. The second format is used when handling data on the OpenStreetMap servers, and isn't intended for use with any other system. The last format is no longer actively supported, as it isn't used on OpenStreetMap's servers any more, and you shouldn't use this format for new databases.

PostGIS (`http://postgis.refractions.net/`) is a version of the open source PostgreSQL database with geospatial extensions that allows you to retrieve geographic data, such as OpenStreetMap, efficiently. You can use PostGIS with many geographic information systems (GIS) and other tools that support the OpenGIS Simple Features Specification for SQL.

You will get the best performance from your system if you run Osmosis on a separate machine from the database, but this isn't essential. What you will need is a very large amounts of disk space available, of the order of hundreds of gigabytes. Many of the most effective ways of using a database with OpenStreetMap data involve storing the entire planet, which needs over 100GB of storage. If you're planning on using a database of OpenStreetMap data in a production system, it will also be worth using high performance storage systems, such as drive arrays. The main OpenStreetMap database is striped across ten drives, but you probably won't need that many.

Installing PostGIS and creating the database

While PostgreSQL itself runs on many different platforms, PostGIS needs to be installed separately, and binary packages aren't available from the creators of PostGIS. If you don't want to compile the software yourself, the easiest option is to use a Linux distribution with pre-built PostgreSQL and PostGIS packages, such as Ubuntu Linux. That's what we'll use in our examples.

If you don't already have Ubuntu Linux running, it's simple to install it by downloading a disc image from `http://www.ubuntu.com/`, and either installing it on a separate machine, or by using a virtual machine host, such as VirtualBox or VMWare. Ubuntu's installation system is easy enough to follow, so we won't cover installing the operating system here. We're using Ubuntu 10.04 Server edition, PostgreSQL 8.4, and PostGIS 1.4 for these examples.

To install PostGIS, and create a database with the spatial extensions enabled, we need to carry out the following steps:

1. Install PostgreSQL.
2. Install PostGIS.
3. Create our new database.
4. Add the PostGIS geospatial extensions to the database.
5. Create the tables for our OpenStreetMap data.

If you're running a desktop version of Linux, open a command shell, or log in if you're using a server version without a graphical front end.

On Ubuntu Linux, we can combine the first two operations into a single command line using its package management tools, as follows:

```
sudo apt-get install postgresql-8.4-postgis postgresql-contrib-8.4
postgis
```

This installs all the software we need at this stage. You will also need to install Osmosis itself at some point, unless you're planning to access the database from a separate machine.

Using Osmosis on the same machine as PostgreSQL

If you're planning to use Osmosis on the same machine as the database, you'll need to add some extra configuration to PostgreSQL, to allow us to access the local database.

We need to edit the file `pg_hba.conf` that contains the access control information for PostgreSQL. This is located in the directory `/etc/postgresql/<version>/main`.

Open this file in an editor, and find a line that reads:

```
local    all        all              ident sameuser
```

Add the following line before the preceding line in the file:

```
local    osmosis    osmosisuser                      trust
```

Save the file, then restart the database server to apply the new configuration:

```
sudo /etc/init.d/postgreql-<version> restart
```

Creating the database

Next, we create a database for our data and a user profile for Osmosis to use. First, we change to the PostgreSQL administrator account:

```
sudo su - postgres
```

You will be prompted for your password. Now, create the database for our data:

```
createdb osmosis
```

The `createdb` utility will create the database. Next, we need to add PL/pgSQL—a procedural extension to SQL—to our database:

```
createlang plpgsql osmosis
```

We now create a user for Osmosis to access the database, as follows:

```
createuser osmosisuser
```

You will be asked for what privileges the user should have. You can answer no to all the questions by pressing *n*.

The next step is to add the PostGIS extensions to the database. A script is supplied to do this, but is stored in different locations for different versions of PostgreSQL. For PostgreSQL 8.3, use the following command:

```
psql -d osmosis -f /usr/share/postgresql-8.3-postgis/lwpostgis.sql
```

Use the following command on PostgreSQL 8.4:

```
psql -d osmosis -f /usr/share/postgresql/8.4/contrib/postgis.sql
```

Next, we need to change some permissions on our database so that Osmosis can access all the tables we've created so far. Start the command-line PostgreSQL client using the following command:

```
psql -d osmosis
```

So far we've used the database superuser account, postgres, to perform all our operations. We need to change the owner of the database and the tables created so far, and make sure our user account has full access to the database.

Enter the following commands into the database client, remembering to include the semicolon at the end of each line:

```
grant all privileges on database osmosis to osmosisuser;
alter table geometry_columns owner to osmosisuser;
alter table spatial_ref_sys owner to osmosisuser;
```

Finally, we should add a password to the osmosisuser account, as follows:

```
alter role osmosisuser password 'openstreetmap';
```

That's most of our work done. Exit the PostgreSQL client using the \q command, and log out of the postgres account.

Now, we create the tables that Osmosis will use to store our data. We do this as osmosisuser so that we don't have any problems with permissions:

```
psql -d osmosis -U osmosisuser -W -f osmosis/script/pgsql_simple_schema_
0.6.sql
```

You will be prompted for the password for osmosisuser, which we set to openstreetmap previously. All being well, you should see the script's output scroll past, and you're then ready to start importing data.

If you ever want to start again from scratch, just re-run the schema creation script, which will delete your existing tables and create new ones.

Adding data to the database

To import an entire planet file into a database, you simply read it in using a read XML task, and write it back again using a `--write-pgsql` task. Assuming you're running Osmosis on the same machine:

```
bzcat planet-100428.osm.bz2 | osmosis --read-xml file=/dev/stdin --write-
pgsql database="osmosis" user="osmosisuser" password="openstreetmap"
host="localhost"
```

All the data in your source is included in the database. You can use tag filters or bounding box/polygon tasks between the input and output if you're only interested in certain features or areas. However, if you want to keep your data in sync with the OpenStreetMap database, it's best for you to keep a full copy of the planet and update that, as updating an extract using the diffs is a far more complicated procedure and can produce unpredictable results.

Reading data from the database

You can read data back out of the database into a pipeline with the `--read-pgsql` task, so to extract the UK from our database, we'd use a command similar to the one we used with the raw planet file earlier in the chapter:

```
osmosis --read-pgsql database="osmosis" user="osmosisuser"
password="openstreetmap" --dataset-dump --bounding-polygon file="united_
kingdom.poly" --write-xml file="united_kingdom.osm.bz2"
```

We need to include a `--dataset-dump` task to convert the output of the `--read-pgsql` task, which is a dataset, to an entity stream that is needed by the rest of the tasks in the pipeline.

However, now that we've got our data in a database, we can use this to our advantage by extracting the bounding box directly from the database, using the `--dataset-bounding-box` task. This uses a database query to retrieve only the bounding box in which we're interested, saving a large amount of processing in Osmosis itself. It also produces an entity stream, so there's no need for a further conversion step:

```
osmosis --read-pgsql database="osmosis" user="osmosisuser"
password="openstreetmap" --dataset-bounding-box left="-11" bottom="49"
right="2" top="61" --write-xml file="britain_ireland.osm.bz2"
```

The bounding box used here includes the whole of Britain and Ireland, with some margin to spare, but the waters around the territory mean we're not including too much extraneous data. However, this only uses a bounding box, so to cut out the UK precisely, we still need to cut out a bounding polygon in Osmosis as well. The overall operation should still be much quicker than cutting the bounding polygon from an XML planet file.

```
osmosis --read-pgsql database="osmosis" user="osmosisuser"
password="openstreetmap" --dataset-bounding-box left="-11" bottom="49"
right="2" top="61" --bounding-polygon file="united_kingdom.poly" --write-
xml file="united_kingdom.osm.bz2"
```

Similarly, you write your subset back to a separate database if you're planning on accessing the data from an application that can read the PostGIS database directly. Assuming you've set up a database using the same steps as our main database, but instead naming it ukextract, you could use the following command:

```
osmosis --read-pgsql database="osmosis" user="osmosisuser"
password="openstreetmap" --dataset-bounding-box left="-11" bottom="49"
right="2" top="61" --bounding-polygon file="united_kingdom.poly" --write-
pgsql database="ukextract" user="osmosisuser" password="openstreetmap"
host="localhost"
```

Applying changes to the database

Just as we patched a planet file earlier in the chapter using diffs files, Osmosis can apply those same changes directly to our database to keep it up-to-date. This should be a much faster operation, as it's not necessary to read in the whole of the data in the database in order to patch it.

To get started, we need an initialized replication folder for Osmosis. Follow the instructions earlier in the chapter to create a set of replication files, and download a starting state file.

Import a downloaded planet file into your database using the commands shown previously. Then, to perform an update, instead of applying the changes in the collected diffs to an entity stream, you simply write them directly to the database using the --write-pgsql-change task:

```
osmosis --read-replication-interval  workingDirectory="replicat
ion" --simplify-change --write-pgsql-change database="osmosis"
user="osmosisuser" password="openstreetmap" host="localhost"
```

Again, this will download all the necessary diff files, combine them into a single set of changes, and apply them to our database. We can re-run this command at any time to keep our database up-to-date.

What you can't reliably do is apply the changes to a subset of the planet file. The diffs on the planet site that Osmosis uses in its replication tasks represent all changes to OpenStreetMap's data for the entire planet and every type of feature. Although you could import a subset of a planet file initially, applying the diffs to your database would quickly introduce unwanted features into the data.

If you need a subset of OpenStreetMap's data that you need to keep up-to-date, one way of achieving it is to keep a full planet database, which you update as shown previously, then regularly re-extract your subset of data into a second database.

Using an auth file to store database credentials

If you're going to be using a database with Osmosis, you can save yourself a lot of typing by storing the username and password you use to access the database in a separate file, called an auth file.

The auth file is a plain text file with names and values of the parameters we normally pass to Osmosis on the command line. Create a file called `dbauth` and add the following to it, changing any details to match those you've used:

```
host=localhost
database=osmosis
user=osmosisuser
password=openstreetmap
dbType=postgresql
```

You can leave out any of the previous values, and specify them on the command line instead, or you can override the values in an auth file on the command line. This allows you to use a single auth file for multiple databases on the same host. Just remember which one you're meant to use!

Now, to access the database from Osmosis, you simply specify the location of the auth file as a parameter to the `--read-pgsql` task:

```
osmosis --read-pgsql authFile="dbauth" --bounding-polygon …
```

You should be able to copy an auth file from one machine to another if you need to.

Other Osmosis tasks

We haven't covered every task that Osmosis can perform. Others, including tasks to generate reports about data, and to deal with legacy file and database formats, are also supported. You generally won't need to use any of these tasks while using the data, but they can be useful when trying to find obscure errors in the data itself. For a full listing of the tasks Osmosis supports, visit `http://wiki.openstreetmap.org/wiki/Osmosis/Detailed_Usage`.

Even if Osmosis can't yet do what you need it to do, you can extend it if you know enough Java. There's more information on writing your own plugins at `http://wiki.openstreetmap.org/wiki/Osmosis/Writing_Plugins`.

Summary

Given enough time, storage, and processing power, Osmosis allows you to extract, adapt, and store OpenStreetMap data in more or less any way you can imagine. If you are working with OpenStreetMap data in country-sized quantities or larger, it should be one of the first tools you use to get the information you want from the raw data.

You should now know how to perform the following tasks with Osmosis:

- Read data from a planet file or a pre-built extract.
- Read data directly from the OpenStreetMap API or XAPI
- Write processed data to a file or database
- Cut out an area of data using a bounding box or a bounding polygon
- Extract a subset of features from a dataset using tag filters
- Keep your source of data up-to-date using replication

If you use large amounts of OpenStreetMap data in an application or to create a map, eventually you will encounter problems that can only realistically be solved using Osmosis. For this reason, it's worth getting to know how Osmosis works, what it's capable of, and how to get it to do what you need. You'll save yourself a lot of time later.

11
OpenStreetMap's Future

OpenStreetMap has achieved a great deal in a very short time, but there's still more to do. We'll explain some of the changes being developed by the coders and mappers working on OSM, and how they'll affect users of the data.

Obviously, the most important task for the project is to increase the quantity and accuracy of the data, but there are changes to the tools used that can make that process easier. There are currently no plans to make large changes to the OpenStreetMap data model or API, and changes to the main editing applications continue at a steady rate.

Changing the OpenStreetMap license

The most significant change to OpenStreetMap in its immediate future isn't technical, but legal. The license used for OpenStreetMap's data is likely to change in the near future, and may have already changed by the time you read this.

At the time of writing, OpenStreetMap uses the Creative Commons Attribution-Share Alike 2.0 license (CC-BY-SA) for its data and map tiles. The intention of using this license is to allow use of the data for any purpose, by anyone, without further approval or permission, while requiring those using the data to make it available to those they distribute it to under the same or similar licensing conditions.

The terms of the license also prevent the data from being combined with an existing proprietary dataset without the resulting dataset being available to the contributors to OpenStreetMap. Some within the OpenStreetMap community feel this is necessary, because while OpenStreetMap's data is accurate and up to date, it's not complete.

Without such a provision in the license for the data, a proprietary geographic database provider could use the parts of OpenStreetMap's data that were more recent than its own, but not let OpenStreetMap have access to its own data and complete the areas no mappers had yet surveyed. The company would be taking advantage of the efforts made by mappers without making any contribution of its own, which supporters of share-alike licensing felt went against the ethos of the project.

However, one issue with using CC-BY-SA for the map data is that it is unclear how far the share-alike component of the license extends. If someone includes an OpenStreetMap map in a brochure, does it mean the whole brochure also has to be CC-BY-SA? While many people believe this isn't the case, nor do they want it to be, the legal position wasn't entirely clear.

At the time OpenStreetMap was set up, CC-BY-SA was a common license for creative works that their creators wanted to share with others and allow their reuse. While any maps created by rendering data from OpenStreetMap are clearly creative works, as the design details of any map rendering were needed to be chosen, the situation for the data itself is less obvious.

What became clear as the project became more successful is that not all jurisdictions give copyright protection to factual information (which OpenStreetMap's data can be considered to be) or recognize the collation of a database as a creative effort. This meant that in some countries it might be legal to ignore the terms of CC-BY-SA and incorporate OpenStreetMap's data into a proprietary database.

The board of the OpenStreetMap Foundation felt that a new license was needed to achieve the same effects as the original CC-BY-SA license, that is, to allow free reuse of the data but prevent its wholesale incorporation in proprietary databases.

Adopting the Open Database License

After a search, the OSMF found that no such license existed, but work on one, called the **Open Database License (ODbL)**, had been started but later abandoned. The OSMF sponsored some work towards the completion of the license, and a project called Open Data Commons, part of the Open Knowledge Foundation, was asked to be a host organization for the new license. The resulting Open Database License (http://www.opendatacommons.org/licenses/odbl/1.0/) has been created for datasets such as OpenStreetMap, and OpenStreetMap is one of the first projects to consider using it. While the license hasn't been written specifically for OpenStreetMap, it was one of the projects considered during its creation. It uses a combination of copyright and contractual law to enforce the aims of the license, and so doesn't appear to suffer the same problems as using CC-BY-SA for data.

The Open Database License also introduces the concept of a produced work, that is, something created using the data in the database. In OpenStreetMap's case, this is any maps rendered from the data. The new license makes clear the difference between the data and produced works, and makes it clear that only the data behind any produced work needs to be shared, not the produced work itself.

The Foundation's members were polled in late 2009 on whether they supported a change of license to ODbL, and of those members that voted, a large majority were in favor. At the time of writing, preparation work was underway on the process of asking contributors to re-license their data, and to change the sign-up terms of the new license. While it's still possible that many existing contributors could refuse to re-license their data under the new license, this isn't expected to happen.

MapCSS—a common stylesheet language for OpenStreetMap

We've seen in this book that there are many tools available to render maps from OpenStreetMap data, but they all have different ways of creating rendering rules that differ to lesser or greater extents. While this allows you to get the most from each rendering application, it also means that creating the same style for different applications means writing new sets of rules for each one.

MapCSS is an effort to produce a unified stylesheet language for OpenStreetMap-based maps that any rendering or editing application can use. It uses the same syntax as the World Wide Web Consortium recommendation Cascading Style Sheets (CSS) — the language used for web stylesheets.

Designers creating rendering rules for OpenStreetMap would no longer need to write separate sets of rules for each rendering application they wanted to use, but instead would write MapCSS stylesheets, which are then either used natively by the application or translated into the renderer's native format.

MapCSS is at a very early stage of development, but the intention is to follow existing standards as much as possible, while still allowing designers to create the maps they want.

MapCSS uses CSS selectors to choose which features to render, and then defines the rendering rules for those features using CSS syntax, as follows:

```
way[highway=trunk]
{
  z-index: 8;
  color: green;
  width: 5;
  casing-color: black;
  casing-width: 7;
}
```

Here the selector chooses any way tagged with `highway=trunk`, colors it green, sets a width of 5, and adds a black casing. Most of this is standard CSS as used to add style to SVG documents, but the `casing-color` and `casing-width` instructions are specific to MapCSS.

As MapCSS will support standard CSS conventions, you can give the same style to two types of features easily, as follows:

```
way[highway=trunk], way[highway=trunk_link]
{
  ...
}
```

Here the multiple selectors are separated by a comma, and you can include as many selectors in this way as you like.

Similarly, MapCSS supports the use of complex selectors, so that to render a way that's part of a long-distance route, you'd use a rule like the following:

```
relation[type=route] way
{
  ...
}
```

Selectors don't have to specify the full tag for a feature, so it should be possible to create a basic set of rules for all ways tagged with `highway=*` using the following selector:

```
way[highway]
{
  ...
}
```

Then add further rules for particular types of road using a more specific selector.

The formatting properties available in MapCSS will vary between renderers, just as it has for web browsers supporting CSS for HTML. A core set of properties, similar to those used when formatting Scalable Vector Graphics using CSS, has been proposed, but not all of the properties used for SVG are appropriate for all renderers, especially where icons and patterns are being used on maps.

MapCSS will also support adding labels and "shields", or block captions to features, although the syntax of doing so is still under development. It's also possible that interactive map renderings, with :hover and :active pseudo-classes providing a way of adding properties to maps when they're viewed online, assume the map viewer is interactive and not using pre-rendered images.

The first application to use MapCSS is Halcyon—the Flash-based rendering engine used in version 2 of Potlatch, the online editor. It can be seen in the following screenshot. At the time of writing, this was a working prototype, but still didn't support all the planned features of MapCSS.

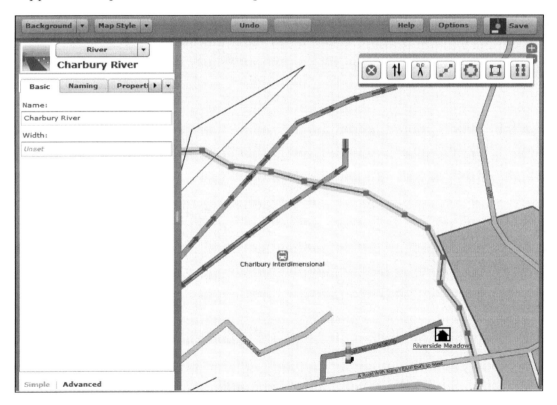

Translation programs that turn MapCSS into the native formats used by Osmarender, Merkaartor, Mapnik, and other renderers may well be the next applications to appear. You can read more about the plans for MapCSS on the OpenStreetMap wiki at `http://wiki.openstreetmap.org/wiki/MapCSS` or `http://bit.ly/mapcss`.

Specialized editing applications

OpenstreetMap's data format is deliberately simple and flexible to allow mappers as much freedom as possible in the way they record the geographic features they encounter. The disadvantage of this approach is that it's more difficult for non-technical users to add information to the data, as for many features you need to know how to draw and tag the primitives to represent the feature you're mapping.

One solution being proposed is to have specialized editors for different types of features, so that motorists can map road-related features, walkers can map footpaths, and cyclists can map cycle paths, each using a tailored interface. The theory is that by not including features in which a particular type of mapper isn't interested, the editing application's user interface can be simplified, increasing usability.

One such editor has already been produced by Cloudmade: the Mapzen POI Collector (`http://mapzen.cloudmade.com/mapzen-poi-collector`) for the Apple iPhone. This small editing application allows users to add points of interest such as shops, cafés, amenities, and other useful places using the iPhone's built-in GPS and OpenStreetMap maps.

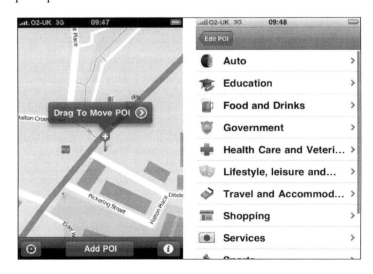

You simply enter the details of the point of interest of your current location, choosing from its built-in categories to find the type of feature you're mapping. There's no facility to add roads or other linear features, but that makes the interface for Mapzen POI collector easy to use even without any prior mapping experience.

The existing desktop editing applications, JOSM and Merkaartor, include customizable presets and rendering rules, so they could be turned into a specialized editor, provided the effort is made to create the necessary files. Potlatch 2 will feature customizable rendering and presets, so it will also be able to provide a similar facility.

One final idea being discussed in the OpenStreetMap community is allowing mappers to edit the properties of a feature, but not its geometry, through a web interface on openstreetmap.org. This would allow people with no mapping experience to fix problems, such as incorrect street names or route references. No work has been done on implementing this feature to date, but it remains a topic of discussion.

Summary

You should now know how to use OpenStreetMap's data for whatever purpose you want. In particular, remember the following points:

- You will never have to pay for OpenStreetMap data. While you can pay for services providing OpenStreetMap data, the raw data will always be free.

- You can, and should, correct any errors or omissions you find in the data.

- You can redistribute OpenStreetMap data to anyone, for any purpose.

- OpenStreetMap is a community-run project, and you can get more information and help by using the community's various communication channels.

- There are several OpenStreetMap editing applications available, depending on your platform, skill level, and connectivity.

- You can create maps from OpenStreetMap data using a number of rendering applications, depending on the level of customization you want and the platform you're running on.

- You can download bulk data in the form of a planet file, or selective datasets from the extended API (XAPI).

- You can produce tailored subsets of OpenStreetMap data using Osmosis to your exact specifications.

- You can use OpenStreetMap data in a traditional Geographic Information System (GIS).

OpenStreetMap passed the point of being a viable project long ago, and has reached a reasonable level of technical stability. While the data itself will change at an increasing rate as OpenStreetMap becomes more popular, the greatest level of software development around OpenStreetMap data is expected to be outside the core infrastructure, as people come up with more and varied ways of using the data.

As already noted, there are no great changes expected in OpenStreetMap's data model, and the work taking place on the editing applications should mean that you'll need to know less about the internals of the OpenStreetMap database to contribute to it in future. There should also be better tools to manage errors in the data, whether accidental or deliberate.

If you have any ideas that you think will improve OpenStreetMap, whether that's the data, the tools surrounding it, or some other aspect of the project, contributions are always welcome. Remember to look in the wiki and mailing list archives for ideas similar to your own, to check the ideas haven't already been discounted, or to find possible collaborators. You'll also need to be willing to put in work on implementing your ideas, as the voluntary, distributed nature of the project means other mappers and developers are unlikely to want to do it for you.

Whatever changes do happen to OpenStreetMap, it won't change the open nature of the project. You will always have access to the data, and you will always be able to use and redistribute it for any purpose. Most of the software we've covered in this book is and will remain open source, so the tools to use the data will always be available. Whatever investment you make in OpenStreetMap, you won't be left with nothing, and that's OpenStreetMap's real future.

Index

CSS classes
  using, for style 167
Cycle map layer 25

D

data
  adding, to database 214
  editing, in JOSM 88, 89
  editing, in Merkaartor 96, 97
  editing, Potlatch used 78
  filtering 128, 129
  filtering, by area 185
  filtering, by associated elements 187, 188
  filtering, by tag 185, 186
  filtering, by user activity 189
  local data, uploading from diffs 205, 206
  reading, from database 214, 215
  sorting 128, 129
database
  changes, applying 215
  creating 211-213
  data, reading 214, 215
  Osmosis, using 210
database credentials
  storing, auth file used 216
data inspection tools, OpenStreetMap 122, 123
data model, OpenStreetMap 53, 54
data model, primitives
  nodes 54-57
  relations 54-60
  ways 54-59
data overlay 25
data processing, Osmosis 193, 194
data streams
  merging 202, 203
  splitting 202, 203
datum 34
defs element 162
dev 28
diff files
  about 174, 176
  downloading 207
  local data, uploading 205, 206
  osmChange format 176
Dilution of Precision. *See* DOP

Docks toolbar 93
DOP 33, 88
drawing features 100-104

E

else rule 169, 170
exporter service 138
extended API. *See* XAPI
Extensible Markup Language. *See* XML
Extensible Stylesheet Language Transform
    (XSLT) 155

F

filters
  about 201
  list 187, 188
  node-key 201
  node-key-value 201
  way-key 201
  way-key-value 201
Flickr
  URL 18
forums 29

G

gates 108
Geofabrik 131, 177
geographic information system. *See* GIS
geonames 21
geotagged images 87
GIS 8, 53
Global Navigation Satellite System (GNSS)
    32
Global Positioning System. *See* GPS
GLONASS 32
GPS 8, 32
GPSBabel
  about 46
  URL 46
GPSDrive 12
GPS Exchange format. *See* GPX
GPS receiver
  about 33
  configuring 38
GPStogo scheme 16

GPS traces
about 34
benefits 35
direct conversion avoiding, reasons 72
roads, mapping 100
using, in Potlatch 80
GPS tracks
adding 145
GPX 46

H

Halcyon 223
Harrow Map
presets, using 114

I

iframe element 140
image files
generating 137, 138
image format
selecting 136, 137
images
files, generating 137, 138
format, selecting 136, 137
loading, into JOSM 87
loading, into Merkaartor 95, 96
Inkscape
maps 159
SVG, editing 158
URL 158
installation, Kosmos 141
installation, Osmarender 157
Internet Relay Chat. *See* **IRC**
IRC
using 28
ITO 128

J

JOSM
about 13, 37, 71, 81, 225
account information, adding 91, 92
authenticating, to OpenStreetMap server 91
data, editing 88, 89
extending, with plugins 92
features 71, 81

images, loading into 87
latest version 81
parking area, drawing 104
presets, used 90
requisites 71
roads, mapping 102
standard view 87
tested version 81
URL 81
user interface 82-87
using, on public computers 82
versus Merkaartor 71
versus Potlatch 71
wireframe view 87

K

keys 68
Kosmos
about 140
bitmap, exporting 151
Console version 152
GPS tracks, adding 145
installing 141
maps, rendering on Windows 140, 141
map tiles, rendering 153, 154
OpenStreetMap data, loading into 143, 144
OpenStreetMap maps, adding 145
project, creating 142, 143
rendering rules, customizing 145-150
Kosmos Console
using 152

L

Landsat information 12
latest version, JOSM 81
legal talk 28
license, OpenStreetMap
changing 219, 220
loops
ways, drawing 105

M

MacPorts
URL 174

U

Ubuntu
URL 211
undocumented tags
finding 115, 116
unsurveyed areas
finding, NoName layer used 127, 128
user account
creating 21
settings 22, 23
user activity
using, for data filter 189
user interface, JOSM 82-87
user interface, Merkaartor 93, 95
user interface, Potlatch 75-78
user segment, NAVSTAR 33
UTF-8 encoding scheme 61

V

video mapping
limitations 37
view
creating 130
views, JOSM
standard 87
wireframe 87
virtual nodes 89
visible attribute 55, 181

W

Walking Papers
URL 51
way-key-value filter 201
way-key filter 201

ways
about 54, 57
drawing, with loops 105
example 58
XML format 58
web page
maps, embedding 138-140
wget utility 174
WGS84 coordinate system 34, 38
wiki-like editing 54
Wikipedia
versus OpenStreetMap 10
Windows
maps rendering, Kosmos used 140, 141
wireframe view, JOSM 87

X

XAPI
about 183
data, filtering by area 185
data, filtering by associated elements 187, 188
data, filtering by tag 185, 186
data, filtering by user activity 189
map query 184
queries 183
query by primitive 184
standard API calls 184
URL 183
XAPI queries 183
XML 173
XML format 55
XMLStarlet
XSL processing 156, 157
XSL processing
XMLStarlet, used 156, 157

Thank you for buying
OpenStreetMap

About Packt Publishing

Packt, pronounced 'packed', published its first book "*Mastering phpMyAdmin for Effective MySQL Management*" in April 2004 and subsequently continued to specialize in publishing highly focused books on specific technologies and solutions.

Our books and publications share the experiences of your fellow IT professionals in adapting and customizing today's systems, applications, and frameworks. Our solution based books give you the knowledge and power to customize the software and technologies you're using to get the job done. Packt books are more specific and less general than the IT books you have seen in the past. Our unique business model allows us to bring you more focused information, giving you more of what you need to know, and less of what you don't.

Packt is a modern, yet unique publishing company, which focuses on producing quality, cutting-edge books for communities of developers, administrators, and newbies alike. For more information, please visit our website: www.packtpub.com.

About Packt Open Source

In 2010, Packt launched two new brands, Packt Open Source and Packt Enterprise, in order to continue its focus on specialization. This book is part of the Packt Open Source brand, home to books published on software built around Open Source licences, and offering information to anybody from advanced developers to budding web designers. The Open Source brand also runs Packt's Open Source Royalty Scheme, by which Packt gives a royalty to each Open Source project about whose software a book is sold.

Writing for Packt

We welcome all inquiries from people who are interested in authoring. Book proposals should be sent to author@packtpub.com. If your book idea is still at an early stage and you would like to discuss it first before writing a formal book proposal, contact us; one of our commissioning editors will get in touch with you.

We're not just looking for published authors; if you have strong technical skills but no writing experience, our experienced editors can help you develop a writing career, or simply get some additional reward for your expertise.

Drupal E-commerce with Ubercart 2.x

ISBN: 978-1-847199-20-1   Paperback: 364 pages

Build, administer, and customize an online store using Drupal with Ubercart

1. Create a powerful e-shop using the award-winning CMS Drupal and the robust e-commerce module Ubercart

2. Create and manage the product catalog and insert products in manual or batch mode

3. Apply SEO (search engine optimization) to your e-shop and adopt turn-key internet marketing techniques

Joomla! E-Commerce with VirtueMart

ISBN: 978-1-847196-74-3   Paperback: 476 pages

Build feature-rich online stores with Joomla! 1.0/1.5 and VirtueMart 1.1.x

1. Build your own e-commerce web site from scratch by adding features step-by-step to an example e-commerce web site

2. Configure the shop, build product catalogues, configure user registration settings for VirtueMart to take orders from around the world

3. Manage customers, orders, and a variety of currencies to provide the best customer service

4. Handle shipping in all situations and deal with sales tax rules

Please check **www.PacktPub.com** for information on our titles

Magento: Beginner's Guide

ISBN: 978-1-847195-94-4   Paperback: 300 pages

Create a dynamic, fully featured, online store with the most powerful open source e-commerce software

1. Step-by-step guide to building your own online store

2. Focuses on the key features of Magento that you must know to get your store up and running

3. Customize the store's appearance to make it uniquely yours

4. Clearly illustrated with screenshots and a working example

Selling Online with Drupal e-Commerce

ISBN: 978-1-847194-06-0   Paperback: 264 pages

Walk through the creation of an online store with Drupal's e-Commerce module

1. Set up a basic Drupal system and plan your shop

2. Set up your shop, and take payments

3. Optimize your site for selling and better reporting

4. Manage and market your site

Please check **www.PacktPub.com** for information on our titles

PHP Web 2.0 Mashup Projects

ISBN: 978-1-847190-88-8   Paperback: 304 pages

Create practical mashups in PHP grabbing and mixing data from Google Maps, Flickr, Amazon, YouTube, MSN Search, Yahoo!, Last.fm, and

1. Expand your website and applications using mashups

2. Gain a thorough understanding of mashup fundamentals

3. Clear, detailed walk-through of the key PHP mashup building technologies

4. Five fully implemented example mashups with full code

jQuery UI 1.7

ISBN: 978-1-847199-72-0   Paperback: 392 pages

Build highly interactive web applications with ready-to-use widgets from the jQuery User Interface library

1. Organize your interfaces with reusable widgets: accordions, date pickers, dialogs, sliders, tabs, and more

2. Enhance the interactivity of your pages by making elements drag-and-droppable, sortable, selectable, and resizable

3. Packed with examples and clear explanations of how to easily design elegant and powerful front-end interfaces for your web applications

4. Revised and targeted at jQuery UI 1.7

Please check **www.PacktPub.com** for information on our titles

Expert Python Programming

ISBN: 978-1-847194-94-7   Paperback: 372 pages

Best practices for designing, coding, and distributing your Python software

1. Learn Python development best practices from an expert, with detailed coverage of naming and coding conventions

2. Apply object-oriented principles, design patterns, and advanced syntax tricks

3. Manage your code with distributed version control

4. Profile and optimize your code

Matplotlib for Python Developers

ISBN: 978-1-847197-90-0   Paperback: 308 pages

Build remarkable publication-quality plots the easy way

1. Create high quality 2D plots by using Matplotlib productively

2. Incremental introduction to Matplotlib, from the ground up to advanced levels

3. Embed Matplotlib in GTK+, Qt, and wxWidgets applications as well as web sites to utilize them in Python applications

4. Deploy Matplotlib in web applications and expose it on the Web using popular web frameworks such as Pylons and Django

Please check **www.PacktPub.com** for information on our titles

Printed in Great Britain by
Amazon.co.uk, Ltd.,
Marston Gate.